Intelligent Systems Reference Library

Volume 243

The aim of this series is to publish a Reference Library, including novel advances and developments in all aspects of Intelligent Systems in an easily accessible and well structured form. The series includes reference works, handbooks, compendia, textbooks, well-structured monographs, dictionaries, and encyclopedias. It contains well integrated knowledge and current information in the field of Intelligent Systems. The series covers the theory, applications, and design methods of Intelligent Systems. Virtually all disciplines such as engineering, computer science, avionics, business, e-commerce, environment, healthcare, physics and life science are included. The list of topics spans all the areas of modern intelligent systems such as: Ambient intelligence, Computational intelligence, Social intelligence, Computational neuroscience, Artificial life, Virtual society, Cognitive systems, DNA and immunity-based systems, e-Learning and teaching, Human-centred computing and Machine ethics, Intelligent control, Intelligent data analysis, Knowledge-based paradigms, Knowledge management, Intelligent agents, Intelligent decision making, Intelligent network security, Interactive entertainment, Learning paradigms, Recommender systems, Robotics and Mechatronics including human-machine teaming, Self-organizing and adaptive systems, Soft computing including Neural systems, Fuzzy systems, Evolutionary computing and the Fusion of these paradigms, Perception and Vision, Web intelligence and Multimedia.

Indexed by SCOPUS, DBLP, zbMATH, SCImago.

All books published in the series are submitted for consideration in Web of Science.

Jair Minoro Abe
Editor

Advances in Applied Logics

Applications of Logic for Philosophy, Mathematics and Information Technology

Springer

Editor
Jair Minoro Abe
Graduate Program in Production
Engineering
Paulista University
São Paulo, Brazil

ISSN 1868-4394 ISSN 1868-4408 (electronic)
Intelligent Systems Reference Library
ISBN 978-3-031-35758-9 ISBN 978-3-031-35759-6 (eBook)
https://doi.org/10.1007/978-3-031-35759-6

This Springer imprint is published by the registered company Springer Nature Switzerland AG
The registered company address is: Gewerbestrasse 11, 6330 Cham, Switzerland

Paper in this product is recyclable.

This book is dedicated to my late father, Tadashi, who laid the foundation for achieving my goals. My deep gratitude and sincere homage.

Preface

It is an honour and a privilege to edit this book in honour of Seiki Akama on his sixtieth birthday.

I met him in Ghent, Belgium, in 1997 at the First World Congress on Paraconsistency.

I did not know about his studies on Paraconsistent Logic, let alone his interest in Annotated Logic. Coincidentally, one of the links of our interests was brokered by the renowned logician R. Sylvan (formerly R. Routley).

The following year, we met in Torun, Poland, at the Memorial Symposium Parainconsistent Logic, Logical Philosophy, Informatics and Mathematics.

Since then, a fruitful collaboration began that lasts until today. It is also worth mentioning that I met K. Nakamatsu in Ghent, who contributed significantly to our research in Annotated Logic.

Seiki Akama is a Japanese computer scientist, Engineer Fujitsu Ltd., Kawasaki, Japan, 1984–1993. Lecturer at Teikyo Heisei University, Ichihara, Japan, since 1993. Doctor of Philosophy, Keio University, Yokohama.

One of the topics that Prof. Akama has been dedicated to is Paraconsistent Logic. Among the initial themes, he studied Nelson's Logic, trying to understand how Nelson's characterization of the negation of the system is structured, which is a paraconsistent logic. It has contributions full of originality.

Recently he has also devoted himself to the topic of Rough Sets, both theoretical and applicability, with the collaboration of Y. Kudo and T. Murai.

He also has texts of a philosophical nature (Philosophy of Logic), showing the scope of his studies more comprehensively.

In addition to his regular studies, Akama is also a musician, the violin being one of his favourite instruments.

On his sixtieth birthday, this book is dedicated to Seiki Akama to contribute to science.

The book consists of the following works:

Chapter 1 summarized Seiki Akama's main contributions. Akama explores the foundations and applications of formal Logic. In particular, he has done much work on non-classical Logic. In addition, he published many books both in Japanese and English. This paper briefly surveys his scientific work.

Chapter 2 by Francisco A. Doria, Carlos Alberto Nunes Cosenza, Luis Claudio Bernardo Moura describes the counterexample function as P = NP, a Busy Beaver-like function as it overtakes all computable functions in its peaks and is also a noncomputable function. They are of interest per se and for possible applications in computer science.

Chapter 3 by Seiki Akama and Jair Minoro Abe writes "On the Choice of Primitives in Tense Logic". As Humberstone pointed out, this can be shown for standard modal logic K. The author shows correct axiomatization Kt. If F and P are primitive, as McArthur did, the resulting system is incomplete for Kripke semantics.

Chapter 4 by João Inácio da Silva Filho, Mauricio Conceição Mario, Dorotéa Vilanova Garcia, Raphael Adamelk Oliveira, Maurício Fontoura Blos, Hyghor Miranda Côrtes, and Jair Minoro Abe makes analogies between the main foundations of Paraconsistent Annotated Logics and the equation of the Logistic Map of Verhulst, who originated the modern Chaos Theory and its variations. This comparison shows that the procedures used in applying the Logistic Map identify the Chaos Theory with paraconsistent analysis based on the fundamental concepts of Paraconsistent Logic, arising significant ways of researching the behaviour and stability of chaotic systems.

Chapter 5 by Arnaldo de Carvalho Jr., João Inácio da Silva Filho, Márcio de Freitas Minicz, Gustavo R. Matuck, Hyghor Miranda Côrtes, Dorotéa Vilanova Garcia, Paulo Marcelo Tasinaffo, and Jair Minoro Abe presents in chapter a new form of the $PANC_L$ in the context of paraconsistent artificial neural networks based on extraction of the contradiction effects between the input and the previous output, called $PANC_{LCTX}$. The results demonstrate that the proposed cell effectively integrates asymptotic mode values with practical applications for the industry in signal analysis, estimation and treatment.

Chapter 6 by Pedro Cabalar, Jorge Fandinno and Luis Fariñas del Cerro considers the definition of a Probabilistic Epistemic Logic (PE) and its non-monotonic extension, called Probabilistic Autoepistemic Equilibrium Logic (PAEE).

Chapter 7 by Yasuo Kudo and Tetsuya Murai is concerned with the Rough set theory, initially proposed by Z. Pawlak. It provides a mathematical basis for set-based approximation of concepts and logical data analysis. In this chapter, the authors review an approach of rough-set-based data analysis.

Chapter 8 by Yotaro Nakayama, Seiki Akama and Tetsuya Murai. A bilattice is an algebraic lattice representing both degrees of truth and epistemic state with the amount of information for a proposition. In the chapter, the authors propose a construction of a bilattice with an approximation space of rough sets. Rough sets are an approximation space for an equivalence relation and are adopted to manage uncertain and inconsistent information. The information system of rough sets can be represented with decision logic, which can be reconstructed with a deduction system based on a bilattice. A pair of rough sets as bilattice elements is discussed, and a deductive system with tableau calculi is constructed.

Chapter 9 by Liliam S. Sakamoto, Jair Minoro Abe, Jonatas S. de Souza, Luiz A. de Lima. Corporations worldwide have a growing problem: orchestrating the organization and understanding structured and unstructured data. This study aims to optimize this analysis by minimizing the level of data loss using the Paraconsistent Evidential Annotated Logic Eτ with exciting results.

Chapter 10 by Samira S. Nascimento, Jair Minoro Abe, Nilton C. F. Teles and Cristina C. de Oliveira. The generation of the knowledge economy adds value to human resources in their behavioural skills favouring the organization. Companies must focus on recruitment and selection processes as employees build skills in a competitive environment. However, the factors involved in the selection process of project management professionals do not consider behavioural skills as a strategy. The study's objective is to simulate a selection of two candidates for a project manager, applying the Paraconsistent Annotated Evidential Logic Eτ, collaborating with the decision-making process.

Chapter 11 by Fábio R. Carvalho presents a decision-making method based on paraconsistent annotated evidential Logic Eτ supported by an expert system. The technique uses the Para-analyzer Algorithm as a basis and is explained in detail with illustrations of how it can be applied.

Chapter 12—Annotated logics and application—an overview—Jair Minoro Abe. The author comments on the annotated Logic, which was born in the late 80s and was the main subject of investigation in his career. Abe is considered one of the foremost pioneers in the applications of paraconsistent Logic. The chapter closes with theoretical and application aspects.

The Editor wishes to thank all contributors for the high contribution level that has come to light, despite the numerous difficulties we have experienced in recent years. The Editor also thanks Prof. Dr. Lakhmi C. Jain for the helpful reception he gave us to this project.

São Paulo, Brazil Jair Minoro Abe

Contents

Contributors

Jair Minoro Abe Graduate Program in Production Engineering, Paulista University, São Paulo, Brazil

Seiki Akama C-Republic, Inc., Asao-ku, Kawasaki, Japan

Maurício Fontoura Blos Laboratory of Applied Paraconsistent Logic, Santa Cecília University—Unisanta, Santos, SP, Brazil

Pedro Cabalar University of Corunna, A Coruña, Spain

Hyghor Miranda Côrtes Laboratory of Applied Paraconsistent Logic, Santa Cecília University—Unisanta, Santos, SP, Brazil

Carlos Alberto Nunes Cosenza LabFuzzy, PEP-COPPE, UFRJ, Rio de Janeiro, Brazil

João Inácio Da Silva Filho Laboratory of Applied Paraconsistent Logic, Santa Cecília University—Unisanta, Santos, SP, Brazil

Arnaldo de Carvalho Jr. Instituto Federal de Educação, Ciência e Tecnologia de São Paulo –IFSP, São Paulo, Brazil;
Laboratory of Applied Paraconsistent Logic, Santa Cecília University, São Paulo, Brazil

Fábio Romeu de Carvalho Universidade Paulista—UNIP, Programa de Doutorado em Engenharia de Produção, São Paulo, SP, CEP, Brasil

Márcio de Freitas Minicz Instituto Tecnológico de Aeronautica (ITA), Division of Computer Science, Marechal Eduardo Gomes Square, São Paulo, Brazil

Luiz Antonio de Lima Paulista University, São Paulo, Brazil

Jonatas Santos de Souza Paulista University, São Paulo, Brazil

Francisco A. Doria LabFuzzy, PEP-COPPE, UFRJ, Rio de Janeiro, Brazil

Jorge Fandinno University of Nebraska at Omaha, Omaha, USA

Luis Fariñas del Cerro IRIT, University of Toulouse, CNRS, Toulouse, France

Luiz R. Forçan Paulista University, São Paulo, Brazil

Dorotéa Vilanova Garcia Laboratory of Applied Paraconsistent Logic, Santa Cecília University—Unisanta, Santos, SP, Brazil

Yasuo Kudo College of Information and Systems, Muroran Institute of Technology, Hokkaido, Japan

Mauricio Conceição Mario Laboratory of Applied Paraconsistent Logic, Santa Cecília University—Unisanta, Santos, SP, Brazil

Gustavo R. Matuck Instituto Tecnológico de Aeronautica (ITA), Division of Computer Science, Marechal Eduardo Gomes Square, São Paulo, Brazil

Luis Claudio Bernardo Moura LabFuzzy, PEP-COPPE, UFRJ, Rio de Janeiro, Brazil

Tetsuya Murai Department of Information Systems Engineering, Chitose Institute of Science and Technology, Hokkaido, Japan;
Chitose Institute of Science and Technology, Chitose, Japan

Irenilza A. Nääs Paulista University, São Paulo, Brazil

Yotaro Nakayama BIPROGY Inc., Koto-ku, Tokyo, Japan

Samira S. Nascimento Paulista University, São Paulo, Brazil

Cristina C. Oliveira Paulista University, São Paulo, Brazil

Raphael Adamelk Oliveira Laboratory of Applied Paraconsistent Logic, Santa Cecília University—Unisanta, Santos, SP, Brazil

Liliam Sayuri Sakamoto Paulista University, São Paulo, Brazil

Paulo Marcelo Tasinaffo Instituto Tecnológico de Aeronautica (ITA), Division of Computer Science, Marechal Eduardo Gomes Square, São Paulo, Brazil

Chapter 1
The Scientific Work of Seiki Akama

Jair Minoro Abe

Contents

Abstract Seiki Akama is a logician who explores foundations and applications of formal logic. In particular, he has done a lot of work on non-classical logics. In addition, he published many books both in Japanese and English. This paper briefly surveys his scientific work.

Keywords Seiki Akama · Non-classical logics · Constructive logic · Paraconsistent logic · Natural language

1.1 Introduction

Seiki Akama (Fig. 1.1) is a logician who explores foundations and applications of formal logic. In particular, he has done a lot of work on non-classical logics including modal, constructive and paraconsistent logics. In addition, he published many books both in Japanese and English for beginners and experts.

This paper briefly surveys his scientific work. In Sect. 1.2, we give his biographical information. Section 1.3 describes his scientific work. As he studies various areas related to formal logic, we address the importance of his work. In Sect. 1.4, we discuss his books, focusing on English books. At the end of the paper, complete information on his theses, papers and books may be found.

J. M. Abe (✉)
Graduate Program in Production Engineering, Paulista University, São Paulo, Brazil
e-mail: jair.abe@docente.unip.br

J. M. Abe (ed.), *Advances in Applied Logics*, Intelligent Systems Reference Library 243,
https://doi.org/10.1007/978-3-031-35759-6_1

Fig. 1.1 Seiki Akama

1.2 Biographical Information

Seiki Akama was born on April 29, 1960 in Tokyo, Japan. His father, Hachiro Akama is a physicist. His younger brother, Yohji Akama is a computer scientist, also studying formal logic. In 1979, he began the course of Bachelor of Department of Industrial Administration of Science University of Tokyo in Japan.

At the time, he was interested in natural language and Artificial Intelligence. He studied formal logic under Prof. Arata Ishimoto who is a specialist of modal and constructive logics and natural language semantics since the 1950.

He first read the Japanese versions of Novikov's [9] and Hilbert and Ackermann's [8] textbook translated by Prof. Ishimoto which introduce the material of Hilbert system of classical logic. Professor Ishimoto also taught Schütte type tableaux for classical logic. Since Akama originally learned formal logic via truth-tables in another course, he was surprised in another aspect of logic.

His main concern was natural language understanding, and he studied formal grammar theory in relation to Chomsky's transformational grammar by himself. Professor Ishimoto suggested him to study Montague semantics, and he read Dowty, Wall and Peters' Introduction to Montague Semantics [7].

He rapidly read the book in a week, and Prof Ishimoto was surprised to him. He also attended the course of Prof. Akira Ikeya, who is a formal semanticist held in Sophia University in Japan. The course also dealt with Dowty et al.'s book, he understood the power of Montague semantics for intensional sentences, like Partee's paradox, in natural languages.

In 1984, Akama finished Bachelor course by submitting B.Sc. thesis. He wrote two theses in English. The first thesis "Definite description in Montague grammar" discusses Russell's attribute and referential uses of definite descriptions within the framework of Montague semantics, and was officially submitted as B.Sc. thesis, as Akama [2].

The second thesis [3] discusses "Constructive predicate logic with strong negation and model theory", which gives a Kripke semantics for Nelson's constructive predicate logic with strong negation N. The paper was later submitted to Notre Dame Journal of Formal Logic in 1985 and published in 1988 [JP2].

After his undergraduate study, he started the work in Fujitsu Ltd. which is one of the major computer companies in the world as a programmer. He was working in the company for nine years. His main job is to develop some programming language systems such as COBOL, FORTRAN and assembly languages.

First, he was a member of the groups for the debugger for COBOL, and later implemented mathematical functions for FORTRAN using an assembly language. In 1992, he published a first Japanese book on computational logics [JB1]. The book was novel and was read by many people. Unfortunately, the book was out of print due to the dissolution of the publisher.

In the period, he was very busy, but continued his research and wrote many papers. He wrote a paper on "Resolution in constructivism" published in Logique et Analysis in 1989 [JB1]. In May 1986, he published a paper "A proposal of modal logic programming" for Canadian Artificial Intelligence Conference at Montreal, Canada as his first conference paper. This was the first journey to a foreign country. In the paper, he proposed a resolution calculus for modal logic S5 in connection with a modal extension of logic programming.

In August 1986, he published two papers "Methodology and verifiability in Montage grammar" and "Situational investigation of presupposition" for COLING which is an international conference on computational linguistics. He presented these papers in Bonn. The former discussed several methodological issues in Montague grammar, and the later a formal treatment of presuppositions with in the framework of Barwise and Perry's situation semantics. Akama was a chance to meet Prof. Asher, and was interested in discourse representation theory.

In 1987, Akama presented a paper on the Frame Problem using Veltman' data logic by the invitation of Prof. F. Brown. Akama discussed with Prof. Brown at Canadian Artificial Intelligence Conference in the last year, At the workshop, he communicated with many researchers including J. McCarthy and C. Schwind.

In 1989, Akama submitted to Ph.D. thesis "Constructive Falsity: Foundations and Their Applications to Computer Science" [4] based on his published journal papers to Department of Administration Engineering of Keio University in Japan and received Ph.D. degree in Engineering in 1990. Professor Ura supervised his work. This was the important step for him.

In 1993, Akama went to Toulouse, France as a guest researcher of Univesite Paul Sabatier to work with Prof. Luis Fariñas del Cerro who is a pioneer for using modal logics for computer science. Akama first sent a letter to Dr. David Pearce in Free University of Berlin in Germany but he obtained no reply. Therefore, he decided to go to Toulouse.

Later, he found that Pearce' invitation was not reached to him by receiving a later call from Pearce. The reason to go to Toulouse was that Fariñas del Cerro investigated various applications of modal logic including modal logic programming language MOLOG. Obviously, modal logics was one of his important research topics. The stay in Toulouse was very valuable for him.

After a stay in Toulouse, he got a job at Teikyo University of Technology in Japan as a Lecturer. He taught many courses in Computer Science. In 1997, Akama published a first English book "Logic, Language and Computation" as a Festschrift for Prof. Ikeya in Kluwer (now Springer). He started the project in 1993, and many famous scholars like Max Cresswell, Richard Routley, Nicholas Asher, Luis Fariñas del Cerro provided contributions to the volume.

In connection with his book, Akama invited Fariñas del Cerro in 1993 and Richard Routley (Richard Sylvan) in 1995 to his university. He discussed many issues with them. They also gave some lectures for his students and co-workers. Akama suggested to Sylvan to study relevance logics for computer science. Akama also wrote a joint paper for a new semantics for intuitionistic logic in his stay for Logique et Analyse [JP7]. The paper was accepted, but this good news was reached after Sylvan's sudden death.

In the 1900s, Akama seriously explored potentials of non-classical logics for computer applications. He encountered Subrahmanian's paraconsistent logic programming around 1990, His system was interesting in that it handle both incomplete and inconsistent information, but Akama felt that it is not satisfactory as a logical system. Then, Akama began to work out underlying logical systems for annotated logic programming.

However, in 1991, Akama found that da Costa et al. [5, 6] published two papers to propose so-called annotated logics in Zeitschrift für mathematische Logik. In fact, their papers treated theoretical foundations for annotated logics. Their papers included a citation of one of the authors Jair Minoro Abe's Ph.D. thesis on the subject [1]. Unfortunately, the thesis was written in Portuguese, and Akama stopped to the work for annotated logics.

In 1997, he presented a paper [CP24] on relevant counterfactuals, later published as [BC9] at the First World Congress on Paraconsistency held in Ghent, Belgium, which was a first international conference on paraconsistent logics. The paper gave a theory of counterfactual logics based on relevant logic B allowing paraconsistency. The work was the result of the discussions with R. Routley.

In the conference, there was a session on annotated logics and Prof. Abe presented some papers. Akama attended the session and provided some questions to Abe. After the session, Akama talked with Abe. and found that he can frequently speak Japanese.

In the conference, Akama also met Japanese computer scientist Kazumi Nakamatsu who also studied annotated logic programming. They initiated joint work for annotated logics, and they published the monograph on annotated logics, "Introduction to Annotated Logics" [EB3] in 2016 by Springer

In 2000, Abe organized LAPTEC (the Congress of Logic Applied Technology) held in Sao Paulo and invited Akama as an invited speaker. Akama presented an invited lecture, and met some Brazilian logicians like da Costa for the first time. Akama also first met Prof. Tetsuya Murai who is a computer scientist working on rough set theory using modal logic.

Later, Murai invited Akama to Hokkaido University, and They proceeded to have a strong cooperation on the study of rough set theory with Prof. Yasuo Kudo who is

Murai's student. Later, Akama, Murai and Kudo published a book "Reasoning with Rough Sets" by Springer.

In 2008–2010, Prof. Sadaaki Miyamoto invited Akama as a Visiting Professor of Tsukuba University of Japan. Since Miyamoto was working on rough sets and multisets, they have done some work on these areas. Since 2006, Akama has worked as an advisor of C-Republic of Tokyo, Japan.

By now, Akama did many joint works with several researchers and continued to study a number of subjects related to formal logic. In addition, for Akama, writing books for experts and beginners is another important activity. We will survey his scientific work in Sect. 1.3 and published books in Sect. 1.4.

1.3 Scientific Work

This section briefly surveys Akama's work. We notice that Akama's areas of research are very wide, including the following fields:

- Natural Language Semantics
- Non-Classical Logics
- Rough Set Theory
- Quantum Computing

Natural language semantics is Akama's first research topic since his undergraduate study. He also started with the research on non-classical logics in the same period. He formally investigated Montague semantics and its rival theories like situation semantics and discourse representation theory. He also considered natural language semantics in the context of computational linguistics.

After meeting Prof. Murai, Akama started to explore foundations of rough set theory related to non-classical logics. He also explored rough set theory for various applications. He published two books on rough set theory with Prof, Murai and Prof. Kudo by Springer [EB5, EB6].

Akama's recent research topic is quantum computing, which is a hot topic in computer science. Because quantum computing has some connection with quantum logic, it is not surprising for him to work on quantum computing. When he was an undergraduate student, he seriously learned quantum mechanics, in particular, matrix mechanics.

Akama also serves as a referee for several journals, conferences and publishers. He refereed papers submitted to famous journals including Notre Dame Journal of Formal Logic, Logic Journal of IGPL, Journal of Logic and Computation, Reports on Mathematical Logic, Information Sciences, etc., His review works for conferences include ISMVL, KES, IASTED, etc. He also reviewed several books submitted to Springer.

1.4 Books

Besides scientific research, Akama wrote so many books. In fact, he published 6 English books and 110 Japanese books. The first English book is the Festschrift for one of his teachers, Prof. Ikeya [EB1]. Akama started the project in 1993. Later, he published the book on Quantum computing, annotated logics, paraconsistent logics and rough set theory. He also published the Festschrift for Prof. Abe [EB4].

For Japanese books, they cover most areas of computer science. His first Japanese book [JB1] concerns computational logic including various non-classical logics and it was written while he worked at Fujitsu Ltd. Firstly, he was not interested in writing books. The people of Publishers asked to do, and he wrote many textbooks for his lectures and other lectures at several universities. The first textbook is for discrete mathematics [JB2].

Since Akama is an expert on programming languages, he published books on major programming languages including FORTRAN, C, C++, etc. In addition. he wrote books on the programming languages related to AI, i.e., Prolog and LISP. He also worked with books for statistics, in particular, R and Excel. Akama published books on his research topics, logic, natural language, computing, and on hot topics, such as quantum computing, DNA computing, databases, Now, Akama has some plans for writing further books on his research like non-classical logics, multi set theory, Artificial Life. In future, this paper will be therefore updated.

Seiki Akama's Works

In the following, we list Akama's Works, which are classified by **Theses**, **Journal Papers**, **Book Chapters** and **Conference Papers**.

Theses

[T1] Akama, S.: Definite description in Montague grammar (English). B.Sc. thesis, Department of Industrial Administration, Science University of Tokyo, Japan (1984)
[T2] Akama, S.: Constructive predicate logic with strong negation and model theory (English). Unpublished B.Sc. thesis, Department of Industrial Administration, Science University of Tokyo, Japan (1984). Later published as [JP2]
[T3] Akama, S.: Constructive falsity: foundations and their applications to Computer Science (English). Ph.D. thesis, Department of Administration Engineering, Keio University, Yokohama, Japan (1989)

Journal Papers

[JP1] Akama, S.: Resolution in constructivism. Logique et Analyse **120**, 385–399 (1987) (published in 1999)
[JP2] Akama, S.: Constructive predicate logic with strong negation and model theory. Notre Dame J. Formal Logic **29**, 18–27 (1988)
[JP3] Akama, S.: On the proof method for constructive falsity. Zeitschrift für mathematische Logik und Grundlagen der Mathematik **34**, 385–392 (1988)
[JP4] Ishikawa, A., Akama, S.: Long-distance dependencies in a logic grammar: SCP. J. Inform. Sci. Eng. **5**, 367–377 (1988)

[JP5] Akama, S.: Subformula semantics for strong negation systems. J. Philos. Logic **19**, 217–226 (1990)
[JP6] Akama, S.: The Gentzen-Kripke construction of the intermediate logic LQ. Notre Dame J. Formal Logic **33**, 148–153 (1992)
[JP7] Akama, S.: Curry's paradox in contractionless constructive logic. J. Philos. Logic **25**, 135–150 (1996)
[JP8] Akama, S., Sylvan, R.: Facts, semantics and intuitionism. Logique et Analyse **147–148**, 227–238 (1995) (published in 1997)
[JP9] Akama, S.: Nelson's paraconsistent logics. Logic Logic. Philos. **7**, 101–115 (1999)
[JP10] Abe, J.M., Akama, S.: Annotated logics Q and ultraproduct. Logique et Analyse **160**, 335–343 (1997) (published in 2000)
[JP11] Akama, S., Nagata, Y.: Infon logic based on constructive logic. Logique et Analyse **194**, 119–136 (2006) (published in 2009)
[JP12] Murai, T., Kudo, Y., Akama, S.: A role of granularity and background knowledge in reasoning processes. Kansei Eng. **6**, 43–48 (2006)
[JP13] Akama, S., Nagata, Y.: Prior's three-valued modal logic Q and its possible applications. J. Adv. Comput. Intell. Intell. Inform. (JACIII) **11**, 105–110 (2007)
[JP14] Akama, S., Nagata, Y., Yamada, C.: A three-valued temporal logic for future contingents. Logique et Analyse **198**, 99–111 (2007) (published in 2009)
[JP15] Akama, S., Miyamoto, S.: Curry and fitch on paradox. Logique et Analyse **203**, 271–283 (2008) (published in 2010)
[JP16] Nakamatsu, K., Abe, J.M., Akama, S.: Transitive reasoning of before-after relation based on bf-EVALPSN. KES **2**, 474–482 (2008)
[JP17] Akama, S., Nagata, Y., Yamada, C.: Three-valued temporal logic Qt and future contingents. Studia Logica **88**, 215–231 (2008)
[JP18] Kudo, Y., Murai, T., Akama, S.: A granularity-based framework of deduction, induction and abduction. Int. J. Approx. Reason. (IJAR) **50**, 1215–1226 (2009)
[JP19] Akama, S.: Negative facts and constructible falsity. Int. J. Reason.-Based Intell. Syst. **1**, 85–91 (2009)
[JP20] Akama, S., Murai, T., Miyamoto, S.: A three-valued modal tense logic for the Master argument. Logique et Analyse **213**, 19–30 (2011) (published 2013)
[JP21] Nakamatsu, K., Abe, J.M., Akama, S.: A logical reasoning system of process before-after relation based on a paraconsistent annotated logic program bf-EVALPSN. Int. J. Reason.-Based Intell. Syst. **15**, 145–163 (2011)
[JP22] Akama, S.: Discursive reasoning in a constructive setting. Int. J. Reason.-Based Intell. Syst. **3**, 88–93 (2011)
[JP23] Akama, S., Abe, J.M., Nakamatsu, K.: Constructive discursive logic with strong negation. Logique et Analyse **215**, 395–408 (2011) (published in 2013)
[JP24] Murai, T., Miyamoto, S., Inuiguchi, M., Akama, S.: Granular hierarchical structures of finite nave subsets and multisets based on free monoids and homomorphisms. Int. J. Reason.-Based Intell. Syst. **4**, 118–128 (2012)
[JP25] Akama, S., Murai, T., Kudo, Y.: Epistemic logic founded on nonignorance. Int. J. Intell. Syst. **28**, 883–891 (2013)

[JP26] Murai, T., Miyamoto, S., Inuiguchi, M., Kudo, Y., Akama, S.: Crisp and fuzzy granular hierarchical structures generated from a free monoid. J. Adv. Comput. Intell. Intell. Inform. (JACIII) **18**, 929–936 (2014)
[JP27] Ubukata, S., Murai, T., Kudo, Y., Akama, S.: Variable neighborhood model for agent control introducing accessibility relations between agents with linear temporal logic. J. Adv. Comput. Intell. Intell. Inform. (JACIII) **18**, 937–945 (2014)
[JP28] Murai, T., Miyamoto, S., Inuiguchi, M., Kudo, Y., Akama, S.: Fuzzy multisets in granular hierarchical structures generated from free monoids. J. Adv. Comput. Intell. Intell. Inform. (JACIII) **19**, 43–50 (2015)
[JP29] Akama, S., Murai, T., Kudo, Y.: Partial and paraconsistent approaches to future contingents in tense logic. Synthese **193**, 3639–3649 (2016)
[JP30] Nakayama, Y., Akama, S., Murai, T.: Four-valued tableau calculi for decision logic of rough set. Procedia Comput. Sci. **126**, 383–391. Proceedings of KES2018, Belgrade (2018)
[JP31] Abe, J.M., Akama, S., Nakamatsu, K., da Silva Filho, J.I.: Some aspects on complementarity and heterodoxy in non-classical logics. Procedia Comput. Sci. **126**, 1253–126. Proceddings of KES2018, Belgrade (2018)
[JP32] Nakayama, Y., Akama, S., Murai, T.: Application of granular reasoning for epistemic situation calculus. J. Jpn. Soc. Fuzzy Theory Intell. Inform. **32**, 768–777 (2020)
[JP33] Nakayama, Y., Akama, S., Murai, T.: Bilattice logic for rough sets. J. Adv. Comput. Intell. Intell. Inform. (JACIII) **24**, 774–784 (2020)

Book Chapters

[BC1] Akama, S., Kawamori, M.: Data semantics in logic programming framework. In: Dahl, V., Saint-Dizier, P. (eds.) Natural Language Understanding and Logic Programming, vol. I, pp. 135–151, North-Holland, Amsterdam (1988)
[BC2] Akama, S., Ishikawa, A.: Semantically constrained parsing and logic programming. In: Abramson, H., Rogers, M.H. (eds.) Meta-Programming in Logic programming, pp. 157–168. MIT Press, Cambridge, MA (1988)
[BC3] Akama, S., Ohnishi, H.: Outline of epistemic knowledge base. In: Ras, Z.W., Zemankova, M., Emrich, M.L. (eds.) Methodologies for Intelligent Systems, vol. 5, pp. 110–117, North-Holland, Amsterdam (1990)
[BC4] Ishikawa, A., Akama, S.: A semantic interface for logic grammars and its applications to DRT. In: Brown, C., Koch, G. (eds.) Natural Language Understanding and Logic Programming, vol. III, pp. 281–292, North-Holland, Amsterdam (1991)
[BC5] Akama, S., Ohnishi, H.: Overview of non-monotonic deduction systems. In: Shi, Z. (ed.) Automated Reasoning, North-Holland, Amsterdam (1992)
[BC6] Akama, S.: A meta-level approach to modal logic programming. In: Orgun, M., Ashcroft, E. (eds.) Intensional Programming, vol. I, pp. 260–272. World Scientific Publishing, Singapore (1996)
[BC7] Akama, S.: Recent issues in logic, language, and computation. In: Akama, S. (ed.) Logic, Language and Computation, pp. 1–26. Kluwer, Dordrecht (1997)
[BC8] Akama, S.: On constructive modality. In: Akama, S. (ed.) Logic, Language and Computation, pp. 143–158. Kluwer, Dordrecht (1997)

[BC9] Akama, S.: Relevant counterfactuals and paraconsistency. In: Batens, D., Mortensen, C., Priest, G., Van Bendegem, J.P. (eds.) Frontiers of Paraconsistent Logic, pp. 1–9. Research Studies Press, Baldock, UK (2000)
[BC10] Akama, S.: The degree of inconsistency in paraconsistent logics. In: Abe, J.M., da Silva Filho, J.I. (eds.) Logic, Artificial Intelligence and Robotics, pp. 13–23. IOS Press, Amsterdam (2001)
[BC11] Akama, S.: Non-classical logics and intelligent systems. In: Nakamatsu, K., Jain, L. (eds.) The Handbook on Reasoning-Based Intelligent Systems, pp. 189–205. World Scientific, Singapore (2013)
[BC12] Akama, S.: Introduction. In: Akama, S. Elements of Quantum Computing, Chap. 1, pp. 1–16. Springer, Heidelberg (2015)
[BC13] Akama, S.: Models of a computer. In: Akama, S. Elements of Quantum Computing, Chap. 2, pp. 17–31. Springer, Heidelberg (2015)
[BC14] Akama, S.: Quantum mechanics. In: Akama, S. Elements of Quantum Computing, Chap. 3, pp. 33–56. Springer, Heidelberg (2015)
[BC15] Akama, S.: Quantum computers. In: Akama, S. Elements of Quantum Computing, Chap. 4, pp. 57–89. Springer, Heidelberg (2015)
[BC16] Akama, S.: Applications of quantum computing. In: Akama, S. Elements of Quantum Computing, Chap. 5, pp. 91–100. Springer, Heidelberg (2015)
[BC17] Akama, S.: Future of quantum computing. In: Akama, S. Elements of Quantum Computing, Chap. 6, pp. 101–118. Springer, Heidelberg (2015)
[BC18] Akama, S.: Glossary. In: Akama, S. Elements of Quantum Computing, pp. 119–122. Springer, Heidelberg (2015)
[BC19] Akama, S.: Index. In: Akama, S. Elements of Quantum Computing, pp. 123–126. Springer, Heidelberg (2015)
[BC20] Abe, J.M., Akama, S., Nakamatsu, K.: Introduction. In: Abe J.M., Akama S., Nakamatsu, K. Introduction to Annotated Logics, Chap. 1, pp. 1–4. Springer, Heidelberg (2015)
[BC21] Abe, J.M., Akama, S., Nakamatsu, K.: Propositional annotated logics $P\tau$. In: Abe, J.M., Akama, S., Nakamatsu, K. Introduction to Annotated Logics, Chap. 2, pp. 5–23. Springer, Heidelberg (2015)
[BC22] Abe, J.M., Akama, S., Nakamatsu, K.: Predicate annotated logics $Q\tau$. Abe, J.M., Akama, S., Nakamatsu, K. Introduction to Annotated Logics, Chap. 3, pp. 25–30. Springer, Heidelberg (2015)
[BC23] Abe, J.M., Akama, S., Nakamatsu, K.: Formal issues. In: Abe, J.M., Akama, S., Nakamatsu, K. Introduction to Annotated Logics, Chap. 4, pp. 31–59. Springer, Heidelberg (2015)
[BC24] Abe, J.M., Akama, S., Nakamatsu, K.: Variants and related systems. In: Abe, J.M., Akama, S., Nakamatsu, K. Introduction to Annotated Logics, Chap. 5, pp. 61–110. Springer, Heidelberg (2015)
[BC25] Abe, J.M., Akama, S., Nakamatsu, K.: Applications. In: Abe, J.M., Akama, S., Nakamatsu, K. Introduction to Annotated Logics, Chap. 6, pp. 111–173. Springer, Heidelberg (2015)

[BC26] Abe, J.M., Akama, S., Nakamatsu, K.: Conclusions. In: Abe, J.M., Akama, S., Nakamatsu, K. Introduction to Annotated Logics, Chap. 7, pp. 175–177. Springer, Heidelberg (2015)
[BC27] Abe, J.M., Akama, S., Nakamatsu, K.: References. In: Abe, J.M., Akama, S., Nakamatsu, K. Introduction to Annotated Logics, pp. 179–185. Springer, Heidelberg (2015)
[BC28] Abe, J.M., Akama, S., Nakamatsu, K.: Index. In: Abe, J.M., Akama, S., Nakamatsu, K. Introduction to Annotated Logics, pp. 187–190. Springer, Heidelberg (2015)
[BC29] Akama, S., Abe, J.M., Nakamatsu, K.: Constructive discursive logic: paraconsistency in constructivism. In: Abe, J.M. (ed.) Paraconsistent Intelligent-Based Systems, pp. 23–38. Springer, Heidelberg (2015)
[BC30] Nakamatsu, K., Abe, J.M., Akama, S.: Paraconsistent annotated logic program EVALPSN and its applications. In: Abe, J.M. (ed.) Paraconsistent Intelligent-Based Systems, pp. 39–85. Springer, Heidelberg (2015)
[BC31] Akama, S., Abe, J.M., Nakamatsu, K.: Annotated logics and intelligent control. In: Nakamatsu, K., Kountchev, R. (ed.) New Approaches in Intelligent Control, pp. 301–335. Springer, Heidelberg (2016)
[BC32] Nakamatsu, K., Abe, J.M., Akama, S.: Paraconsistent annotated logic program EVALPSN and its applications to intelligent control. In: Nakamatsu, K., Kountchev, R. (ed.) New Approaches in Intelligent Control, pp. 337–401. Springer, Heidelberg (2016)
[BC33] Akama, S.: Introduction. In: Akama, S. (ed.) Towards Paraconsistent Engineering, Chap. 1, pp. 1–5. Springer, Heidelberg (2016)
[BC34] Akama, S., da Costa, N.C.A.: Why paraconsistent logics? In: Akama, S. (ed.) Towards Paraconsistent Engineering, Chap. 2, pp. 7–24. Springer, Heidelberg (2016)
[BC35] Akama, S.: A survey of annotated logics. In: Akama, S. (ed.) Towards Paraconsistent Engineering, Chap. 5, pp. 49–76. Springer, Heidelberg (2016)
[BC36] Nakamatsu, K., Akama, S.: Programming with annotated logics. In: Akama, S. (ed.) Towards Paraconsistent Engineering, Chap. 7, pp. 103–164. Springer, Heidelberg (2016)
[BC37] Kudo, Y., Murai, T., Akama, S.: A review on rough sets and possible world semantics for modal logics. In: Akama, S. (ed.) Towards Paraconsistent Engineering, Chap. 8, pp. 165–177. Springer, Heidelberg (2016)
[BC38] Murai, T., Kudo, Y., Akama, S.: Paraconsistency, Chellas's conditional logics, and association rules. In: Akama, S. (ed.) Towards Paraconsistent Engineering, Chap. 9, pp. 179–96. Springer, Heidelberg (2016)
[BC39] Akama, S.: Jair Minoro Abe on paraconsistent engineering. In: Akama, S. (ed.) Towards Paraconsistent Engineering, Chap. 12, pp. 227–233. Springer, Heidelberg (2016)
[BC40] Akama, S., Murai, T., Kudo, Y.: Introduction. In: Akama, S., Murai, T., Kudo, Y. Reasoning with Rough Sets, Chap. 1, pp. 1–6. Springer, Heidelberg (2018)

[BC41] Akama, S., Murai, T., Kudo, Y.: Rough set theory. In: Akama, S., Murai, T., Kudo, Y. Reasoning with Rough Sets, Chap. 2, pp. 7–50. Springer, Heidelberg (2018)
[BC42] Akama, S., Murai, T., Kudo, Y.: Rough set theory. In: Akama, S., Murai, T., Kudo, Y. Reasoning with Rough Sets, Chap. 3, pp. 51–84. Springer, Heidelberg (2018)
[BC43] Akama, S., Murai, T., Kudo, Y.: Logical characterizations of rough sets. In: Akama, S., Murai, T., Kudo, Y. Reasoning with Rough Sets, Chap. 4, pp. 85–125. Springer, Heidelberg (2018)
[BC44] Akama, S., Murai, T., Kudo, Y.: A granularity-based framework of reasoning. In: Akama, S., Murai, T., Kudo, Y. Reasoning with Rough Sets, Chap. 5, pp. 127–181. Springer, Heidelberg (2018)
[BC45] Akama, S., Murai, T., Kudo, Y.: Conclusions. In: Akama, S., Murai, T., Kudo, Y. Reasoning with Rough Sets, Chap. 6, pp. 183–186. Springer, Heidelberg (2018)
[BC46] Akama, S., Murai, T., Kudo, Y.: References. In: Akama, S., Murai, T., Kudo, Y. Reasoning with Rough Sets, pp. 187–196. Springer, Heidelberg (2018)
[BC47] Akama, S., Murai, T., Kudo, Y.: Index. In: Akama, S., Murai, T., Kudo, Y. Reasoning with Rough Sets, pp. 197–201. Springer, Heidelberg (2018)
[BC48] Akama, S., Murai, T., Kudo, Y.: Introduction. In: Akama, S., Murai, T., Kudo, Y. Topics in Rough Set Theory, Chap. 1, pp. 1–5. Springer, Heidelberg (2020)
[BC49] Akama, S., Murai, T., Kudo, Y.: Overview of rough set theory. In: Akama, S., Murai, T., Kudo, Y. Topics in Rough Set Theory, Chap. 2, pp. 7–60, Springer, Heidelberg (2020)
[BC50] Akama, S., Murai, T., Kudo, Y.: Object reduction in rough set theory. In: Akama, S., Murai, T., Kudo, Y. Topics in Rough Set Theory, Chap. 3, pp. 61–70, Springer, Heidelberg (2020)
[BC51] Akama, S., Murai, T., Kudo, Y.: Recommendation method for direct setting of preference patterns based on interrelation mining. In: Akama, S., Murai, T., Kudo, Y. Topics in Rough Set Theory, Chap. 4, pp. 71–79. Springer, Heidelberg (2020)
[BC52] Akama, S., Murai, T., Kudo, Y.: Rough-set-based interrelationship mining for incomplete decision tables. In: Akama, S., Murai, T., Kudo, Y. Topics in Rough Set Theory, Chap. 5, pp. 81–99. Springer, Heidelberg (2020)
[BC53] Akama, S., Murai, T., Kudo, Y.: A parallel computation method for heuristic attribute reduction using reduced decision tables. In: Akama, S., Murai, T., Kudo, Y. Topics in Rough Set Theory, Chap. 6, pp. 101–111. Springer, Heidelberg (2020)
[BC54] Akama, S., Murai, T., Kudo, Y.: Heuristic algorithm for attribute reduction based on classification ability by condition attributes. In: Akama, S., Murai, T., Kudo, Y. Topics in Rough Set Theory, Chap. 7, pp. 113–127. Springer, Heidelberg (2020)
[BC55] Akama, S., Murai, T., Kudo, Y.: An evaluation method of relative reducts based on roughness of partitions. In: Akama, S., Murai, T., Kudo, Y. Topics in Rough Set Theory, Chap. 8, pp. 129–140. Springer, Heidelberg (2020)
[BC56] Akama, S., Murai, T., Kudo, Y.: Neighbor selection for user-based collaborative filtering using covering-based rough sets. In: Akama, S., Murai, T., Kudo, Y. Topics in Rough Set Theory, Chap. 9, pp. 141–159. Springer, Heidelberg (2020)

[BC57] Akama, S., Murai, T., Kudo, Y.; Granular computing and Aristotlefs categorical syllogism. In: Akama, S., Murai, T., Kudo, Y. Topics in Rough Set Theory, Chap. 10, pp. 161–172. Springer, Heidelberg (2020)
[BC58] Akama, S., Murai, T., Kudo, Y.: A modal characterization of visibility and focus in granular reasoning. In: Akama, S., Murai, T., Kudo, Y. Topics in Rough Set Theory, Chap. 11, pp. 173–185. Springer, Heidelberg (2020)
[BC59] Akama, S., Murai, T., Kudo, Y.: Directions for future work in rough set theory. In: Akama, S., Murai, T., Kudo, Y. Topics in Rough Set Theory, Chap. 12, pp. 187–198. Springer, Heidelberg (2020)
[BC60] Akama, S., Murai, T., Kudo, Y.: Indexs. In: Akama, S., Murai, T., Kudo, Y. Topics in Rough Set Theory, pp. 199–201. Springer, Heidelberg (2020)

Conference Papers

[CP1] Akama, S.: A proposal of modal logic programming. In: Proceedings of the 6th Canadian Artificial Intelligence Conference, pp. 99–102, Montreal, Canada (1986)
[CP2] Akama, S.: Methodology and verifiability in Montague grammar. In: Proceedings of COLING'86, pp. 88–90, Bonn, West Germany, 1996.
[CP3] Akama, S., Kawamori, M.: Situational investigation of presupposition. In: Proceedings of COLING'86, pp. 174–176, Bonn, West Germany (1986)
[CP4] Akama, S.: Presupposition and frame problem in knowledge bases. In: Proceedings of AAAI Workshop the Frame Problem in AI, pp. 193–203, Lawrence, USA. Morgan Kaufmann (1987)
[CP5] Akama, S., Ishikawa, A.: Semantically constrained parsing in Prolog. In: Lloyd, J. (ed.) Proceedings of the Workshop on Meta-programming in Logic Programming, pp. 121–132, Bristol, UK (1988)
[CP6] Akama, S.: Meta-logic programming for non-monotonic reasoning . In: Proceedings of the 2nd International Symposium on Artificial Intelligence, Monterrey, Mexico (1989)
[CP7] Akama, S.: The rationalist view of modal logic programming. In: Proceedings of the Conference on Fuzzy Logic Programming, pp. 203–211, Ohio, USA. ACM Press (1989)
[CP8] Akama, S.: Semantical considerations on constructive logic with strong negation. In: Proceddings of the Soviet–Japan Symposium on Lesniewski's Ontology and Its Applications, Moscow (1989)
[CP9] Akama, S., Ishikawa, A.: A semantic interface for logic grammars. In: Proceedings of the Seoul International Conference on newline Natural Language Processing, pp. 47–56, Seoul (1990)
[CP10] Akama, S.: Amalgamated logic programming and non-monotonic reasoning. In: Proceedings of the 6th International Symposium on Methodologies for Intelligent Systems, pp. 450–458. Springer, Charlotte (1990)
[CP11] Akama, S., Ohnishi, H.: Metamathematical foundations of non-monotonic reasoning. In: Proceedingsof International Workshop on Automated Reasoning, Beijing (1992)
[CP12] Ohnishi, H., Akama, S.: Indexed knowledge in epistemic logic programming. In: Proceedings of the 3rd Workshop on Meta-Programming in Logic, Uppsala (1992)

[CP13] Akama, S., Ohnishi, H.: Implications in vivid logic. In: Proceedings of ISMIS'93, Poster Session, Torondenheim (1993)
[CP14] Ohnishi, H., Akama, S.: Intentional contexts and common-knowledge. In: Proceedings of the 1st Pacific Asia Conference on Formal and Computational Linguistics, pp. 234–243, Taipei (1993)
[CP15] Akama, S.: A proof system for useful three-valued logics. In: Proceedings of the Japan-CIS Symposium on Knowledge-Based Software Engineering, Pereslavl-Zalesskii (1994)
[CP16] Akama, S., Nakayama, Y.: Consequence relations in DRT. In: Proceedings of COLING'94, pp. 1114–1117, Kyoto (1994)
[CP17] Akama, S., Nakayama, Y.: A three-valued semantics for discourse representations. In: Proceedings of 25th International Symposium on Multiple-Valued Logic (ISMVL'95), pp. 123–128, Bloomington (1995)
[CP18] Akama, S.: Three-valued constructive logic and logic programs. In: Proceedings of the 25th International Symposium on Multiple-Valued Logic (ISMVL'95), pp. 276–281, Bloomington (1995)
[CP19] Akama, S.: A meta-level approach to modal logic programming. In: Proceedings of the 8th International Symposium on Languages for Intensional Programming, pp. 161–168, Sydney (1995)
[CP20] Akama, S., Kobayashi, M.: A labelled deductive approach to DRT I. In: Proceedings of the 5th International Workshop on Natural Language Understanding and Logic Programming, pp. 23–37, Lisbon (1995)
[CP21] Akama, S.: Formalizing implicatures in annotated logic. In: Proceedings of the 5th International Workshop on Natural Language Understanding and Logic Programming, pp. 237–247, Lisbon (1995)
[CP22] Akama, S.: Tableaux for logic programming with strong negation. In: Proceedings of the International Conference on Analytic Tableaux and Related Methods, Pont-a-Mousson, pp. 31–42. Springer, Berlin (1997)
[CP23] Akama, S.: A proof method for the six-valued logic for incomplete information. In: Proceedings of the 27th International Symposium on Multiple-Valued Logic (ISMVL'97), pp. 276–281, Antigonish (1997)
[CP24] Akama, S.: Relevant counterfactuals. In: Proceedings of the 1st World Congress on Paraconsistency, Ghent (1997)
[CP25] Akama, S., Abe, J.M.: Many-valued and annotated modal logics. In: Proceedings of the 28th International Symposium on Multiple-Valued Logic (ISMVL'98), pp. 114–119, Fukuoka (1998)
[CP26] Akama, S.: Nelson's paraconsistent logics. In: Proceedings of Jaskowski's Memorial Symposium, Torun (1998)
[CP27] Akama, S.: A labelled deductive approach to DRT II: attitudes, DRSs and labelled deduction. In: Proceedings of the 1st International Workshop on Labelled Deduction, Freiburg (1998)
[CP28] Akama, S., Abe, J.M.: Natural deduction and general annotated logics. In: Proceedings of the 1st International Workshop on Labelled Deduction, Freiburg (1998)

[CP29] Abe, J.M., Akama, S.: A logical system for reasoning with fuzziness and inconsistencies in distributed systems. In: Proceedings of IASTED, pp. 221–225, Honolulu (1999)
[CP30] Akama, S., Abe, J.M.: Fuzzy annotated logics. In: Proceedings of IPMU'2000, pp. 504–508, Madrid, Spain (2000)
[CP31] Akama, S.: Labelled deduction and dynamic semantics for natural language. In: Proceedings of ASC'2001, pp. 307–312, Cancun, Mexico (2001)
[CP32] Akama, S., Abe, J.M.: Paraconsistent logics viewed as a foundation for data warehouse. In: Abe, J.M., da Silva Filho, J.I. (eds.) Advances in Logic, Artificial Intelligence and Robotics, pp. 96–103. IOS Press, Sao Paulo, Brazil (2002)
[CP33] Akama, S., Abe, J.M., Murai, T.: On the relation of fuzzy and annotated logics. In: Proceedings of ASC'2003, pp. 46–51, Banff, Canada (2003)
[CP34] Akama, S., Abe, J.M., Murai, T.: A tableau formulation of annotated logics. In: Cialdea Mayer, M., Pirri, F. (eds.) Proceedings of TABLEAUX'2003, 1–13, Rome, Italy (2003)
[CP35] Murai, T., Akama, S., Kudo, Y.: Rough and fuzzy sets from a point of view of propositional annotated modal logic. Part 1. In: Proceedings of IFSA'2003, pp. 241–244, Istanbul, Turkey (2003)
[CP36] Akama, S, Nagata, Y.: On Prior's three-valued modal logic Q. In: Proceedings of ISMVL 2005, pp. 14–19, Calgary, Canada (2005)
[CP37] Akama, S, Nagata, Y.: Constructive logic and situation theory. In: Nakamatsu, K., Abe, J.M. (eds.) Advances in Logic Based Intelligent Systems. Proceedings of LAPTEC 2005, pp. 1–8. IOS Press, Amsterdam (2005)
[CP38] Akama, S., Murai, T.: Rough set semantics for three-valued logics. In: Nakamatsu, K., Abe, J.M. (eds.) Advances in Logic Based Intelligent Systems: Proceedings of LAPTEC 2005, pp. 242–247. IOS Press, Amsterdam (2005)
[CP39] Murai, T., Kudo, Y., Akama, S., Abe, J.M.: Paraconsistency and paracompleteness in Chellas's conditional logic. In: Nakamatsu, K., Abe, J.M. (eds.) Advances in Logic Based Intelligent Systems; Proceedings of LAPTEC 2005, pp. 248–255. IOS Press, Amsterdam (2005)
[CP40] Nakamatsu, K., Akama, S., Abe, J.M.: An intelligent safety verification based on a paraconsistent logic program. In: Proceedings of KES 2005, pp. 708–715. Springer, Heidelberg (2005)
[CP41] Abe, J.M., Nakamatsu, K., Akama, S.: Non-alethic reasoning in distributed systems. In: Proceedings of KES 2005, pp. 724–731. Springer, Heidelberg (2005)
[CP42] Akama, S., Nakamatsu, K., Abe, J.M.: A natural deduction system for annotated predicate logic. In: Knowledge-Based Intelligent Information and Engineering Systems: Proceedings of KES 2007—WIRN 2007, Part II, pp. 861–868. Lecture Notes on Artificial Intelligence, vol. 4693. Springer, Berlin
[CP43] Abe, J.M., Akama, S., Nakamatsu, K.: Monadic Curry algebras $Q\tau$. In: Knowledge-Based Intelligent Information and Engineering Systems: Proceedings of KES 2007—WIRN 2007, Part II, pp. 893–900. Lecture Notes on Artificial Intelligence, vol. 4693. Springer, Berlin (2007)
[CP44] Nakamatsu, K., Abe, J.M., Akama, S.: An intelligent coordinated traffic signal control based on EVALPSN. In: Knowledge-Based Intelligent Information

and Engineering Systems: Proceedings of KES 2007—WIRN 2007, Part II, pp. 869–876. Lecture Notes on Artificial Intelligence, vol. 4693. Springer, Berlin (2007)
[CP45] Kudo, Y., Murai, T., Akama, S.: A unified formulation of deduction, induction and abduction using granularity based on VPRS models and measurebased semantics for modal logics. In: Huynh, V.-N., et al. (eds.) Proceedings of the International Workshop on Interval/Probabilistic Uncertainty and Non-Classical Logics (UncLog'08), Ishikawa, Japan, pp. 280–290. Springer, Heidelberg (2008)
[CP46] Abe, J.M., Nakamatsu, K., Akama, S.: Two Applications of paraconsistent logical controller. In: Tsihrintzis, G.A., et al. (eds.) New Directions in Intelligent Interactive Multimedia 2008, pp. 249–254. Springer, Heidelberg (2008)
[CP47] Nakamatsu, K., Abe, J.M., Akama, S.: Paraconsistent before-after relation reasoning based on EVALPSN. In: Tsihrintzis, G.A., et al. (eds.) New Directions in Intelligent Interactive Multimedia 2008, pp. 265–274. Springer, Heidelberg (2008)
[CP48] Akama, S., Nakamatsu, K., Abe, J.M.: Constructive logic and the sorites paradox. In: Tsihrintzis, G.A., et al. (eds.) New Directions in Intelligent Interactive Multimedia 2008, pp. 285–292. Springer, Heidelberg (2008)
[CP49] Abe, J.M., Nakamatsu, K. Akama, S.: An algebraic version of the monadic system C1. In: Nakamatsu, et al. (eds.) New Advances in Intelligent Decision Technologies, Proceedings of the 1st International Symposium IDT 2009, pp. 341–349. Springer, Heidelberg (2009)
[CP50] Akama, S., Nakamatsu, K., Abe, J.M.: Some three-valued temporal logic for future contingents. In: Nakamatsu, et al. (eds.) New Advances in Intelligent Decision Technologies, Proceedings of the 1st International Symposium IDT 2009, pp. 351–361. Springer, Heidelberg (2009)
[CP51] Nakamatsu, K., Imai, T., Abe, J.M., Akama, S.: Introduction to plausible reasoning based on EVALPSN. In: Nakamatsu, et al. (eds.) New Advances in Intelligent Decision Technologies, Proceedings of the 1st International Symposium IDT 2009, pp. 363–372. Springer, Heidelberg (2009)
[CP52] Akama, S.: Time and many-valuedness: applications of three-valued temporal logic. In: Proceedingsof the 18th International Workshop on Post-Binary ULSI Systems (ULSI 2009), pp. 66–70. Okinawa, Japan, May 2009
[CP53] Akama, S., Murai, T., Kudo, Y.: Uncertainty in future: a paraconsistent approach. In: Huynh, V.H. (eds.) Integrated Management and Applications, Advances in Intelligent and Soft Computing 68, Proceedings of IUM 2010, pp. 335–342. Springer, Berlin (2010)
[CP54] Murai, T., Ubukata, S., Kudo, Y., Akama, S., Miyamoto, S.: Granularity and approximation in sequences, multisets and sets in the framework of Kripke semantics. In: Huynh, V.H. (eds.) Integrated Management and Applications, Advances in Intelligent and Soft Computing, vol. 68, Proceedings of IUM 2010, pp. 329–334. Springer, Berlin (2010)
[CP55] Abe, J.M., Nakamatsu, K., Akama, S.: Monadic curry system N_1^*, R. In: Setchi, Jordanov, I., Howlett, R.J., Jain, L.C. (eds.) Knowledge-Based and Intelligent Information and Engineering: Proceedings of KES 2010, pp. 143–153. LNCS, vol. 6278. Springer, Berlin (2010)

[CP56] Abe, J.M., Lopes, H.F.S., Nakamatsu, K., Akama, S.: Paraconsistent artificial neural networks and EEG analysis. In: Setchi, R., Jordanov, I., Howlett, R.J., Jain, L.C. (eds.) Knowledge-Based and Intelligent Information and Engineering, Proceedings of KES 2010, pp. 164–173. LNCS, vol. 6278. Springer, Berlin (2010)
[CP57] Akama, S., Nakamatsu, K., Abe, J.M.: Constructive discursive reasoning. In: Setchi, R., Jordanov, I., Howlett, R.J., Jain, L.C. (eds.) Knowledge-Based and Intelligent Information and Engineering. Proceedings of KES 2010, pp. 200–206. LNCS, vol. 6278. Springer, Berlin (2010)
[CP58] Nakamatsu, K., Abe, J.M., Akama, S., Kountchev, R.: Introduction to intelligent elevator control based on EVALPSN. In: Setchi, R., Jordanov, I., Howlett, R.J., Jain, L.C. (eds.) Knowledge-Based and Intelligent Information and Engineering, Proceedings of KES 2010, pp. 133–142. LNCS, vol. 6278, Springer, Berlin (2010)
[CP59] Akama, S, Nagata, Y.: A three-valued approach to the master argument. In: Proceedings of ISMVL 2011, pp. 44–49 (2011)
[CP60] Abe, J.M., Lopes, H.F.S., Nakamatsu, K., Akama, S.: Applications of paraconsistent artificial neural networks in EEG. In: Proeedings of ICCCI, pp. 82–92 (2011)
[CP61] Murai, T., Miyamoto, S., Inuiguchi, M., Kudo, Y., Akama, S.: Fuzzy multisets in granular hierarchical structures generated from free monoids. In: Proceedings of MDAI 2013, pp. 248–259 (2013)
[CP62] Akama, S., Murai, T., Kudo, Y.: Heyting-Brouwer rough set logic. In: Huynh, V.N., et al. (eds.) Knowledge and Systems Engineering. Proceedings of KSE 2013, vol. 2, Hanoi, Vietnam, pp. 135–145. Springer, Heidelberg (2013)
[CP63] Akama, S., Murai, T., Kudo, Y.: Bi-superintuitionistic logics for rough sets. In: Proceedings of GrC 2013, pp. 10–15 (2013)
[CP64] Akama, S., Murai, T., Kudo, Y.: Da Costa logics and vagueness. In: Proceedings of GrC 2014, pp. 1–6 (2014)
[CP65] Akama, S., Abe, J.M., Nakamatsu, K.: Contingent information: a fourvalued approach. In: Proceedings of KSE 2014, pp. 209–217 (2014)
[CP66] Tanaka, T., Murai, T., Kudo, Y., Akama, S.: Empty-string of the false value in crisp and fuzzy granular hierarchical structures. In: Proceedings of SCIS & ISIS 2014, pp. 993–997 (2014)
[CP67] Akama, S., Abe, J.M., Nakamatsu, K.: Evidential reasoning in annotated logics. In: Proceedings of IIAI-AAI 2015, pp. 28–33 (2015)
[CP68] Abe, J.M., Nakamatsu, K., Akama, S., da Silva Filho, J.I.: Propositional algebra P1. In: Proceedings of KES-IDT, pp. 1–104 (2015)
[CP69] Abe, J.M., Nakamatsu, K., Akama, S., da Silva Filho, J.I.: The importance of paraconsistency and paracompleteness in intelligent systems. In: Proceedings of KES-IDT, no. 2, pp. 196–205 (2017)
[CP69] Abe, J.M., Nakamatsu, K., Akama, S., Ahrary, A.: Handling paraconsistency and paracompleteness in robotics. In: Proceedings of INISTA 2018, pp. 1–7 (2018)
[CP70] Abe, J.M., Akama, S., Nakamatsu, K., da Silva Filho, J.I.: Some aspects on complementarity and heterodoxy in non-classical logics. In: Proceedings of KES 2018, pp. 1253–1260 (2018)

[CP71] Nakayama, Y., Akama, S., Murai, T.: Four-valued tableau calculi for decision logic of rough set. In: Proceedings of KES 2018, pp. 383–392 (2018)
[CP72] Nakayama, Y., Akama, S., Murai, T.: Four-valued semantics for granular reasoning towards frame problem. In: Proceedings of SCIS/ISIS 2018, pp. 37–42 (2018)
[CP73] Akama, S.: Big Data and AI. In: Proceedings of CIF 19 (2019)
[CP74] Nakayama, Y., Akama, S., Murai: Rough set logic for Kleene's three valued logic. In: Proceedings of SCIS/ISIS 2020, pp. 1–5 (2020)

Seiki Akama's Books

We list Seiki Akama's books below. He published many books both in English and Japanese.

English Books

[EB1] Akama, S. (ed.) Logic, Language and Computation. Kluwer, Dordrecht (1997)
[EB2] Akama, S.: Elements of Quantum Computing. Springer, Heidelberg (2015)
[EB3] Abe, J.M., Akama, S., Nakamatsu, K.: Introduction to Annotated Logics. Springer, Heidelberg (2015)
[EB4] Akama, S. (ed.) Towards Paraconsistent Engineering. Springer, Heidelberg (2016)
[EB5] Akama, S., Murai, T., Kudo, Y.: Reasoning with Rough Sets. Springer, Heidelberg (2018)
[EB6] Akama, S., Kudo, Y., Murai, T.: Topics in Rough Set Theory. Springer, Heidelberg (2020)

Japanese Books

[JB1] Akama, S.: Introduction to Computational Logic (Japanese). Keigaku, Tokyo (1992)
[JB2] Akama, S.: Elements of Discrete Mathematics (Japanese). Corona, Tokyo (1996)
[JB3] Akama, S., Hirasawa, K.: Programming with FORTRAN (Japanese). Corona, Tokyo (1996)
[JB4] Akama, S.: Introductory C Language (Japanese). Sugiyama, Tokyo (1997)
[JB5] Akama, S.: Beginning Visual Basic (Japanese). Jikkyo, Tokyo (1997)
[JB6] Akama, S.: Basic Knowledge in Computer Age (Japanese). Corona, Tokyo (1998)
[JB7] Akama, S.: Natural Languages, Semantics and Logic (Japanese). Kyoritsu, Tokyo (1998)
[JB8] Akama, S.: Introduction to Object-Oriented in C++ (Japanese). Kogakutosho, Tokyo (1998)
[JB9] Akama, S.: Introduction to Programming with Visual Basic (Japanese). Maruzen, Tokyo (1999)
[JB10] Akama, S.: Object-Oriented Programming in Java 2 (Japanese). Kyoritsu, Tokyo (1999)

[JB11] Akama, S.: Introduction to Programming in Java (Japanese). Corona, Tokyo (1999)
[JB12] Akama, S.: Numerical Computation in Java 2 (Japanese). Gihodo, Tokyo (1999)
[JB13] Akama, S.: Introduction to Maple (Japanese). Kyoritsu, Tokyo (2000)
[JB14] Akama, S.: Data Processing in Excel (Japanese). Muisuri, Tokyo (2000)
[JB15] Akama, S.: Foundations of Artificial Intelligence (Japanese). Denkishoin, Tokyo (2000)
[JB16] Akama, S.: Elements of Linear Algebras (Japanese). Makishoten, Tokyo (2001)
[JB17] Akama, S.: Principles of Databases (Japanese). Gihodo, Tokyo (2001)
[JB18] Akama, S.: Introduction to MuPAD (Japanese). Springer, Tokyo (2001)
[JB19] Akama, S.: Introduction to Java with Visual J++ (Japanese). Kogakusha, Tokyo (2002)
[JB20] Akama, S.: Basic Knowledge on Information Sciences (Japanese). Gihodo, Tokyo (2002)
[JB21] Akama, S.: Basic Knowledge on Java Programming, Gihodo, Tokyo (2002)
[JB22] Akama, S.: MATLAB Programming Book (Japanese). Syuuwa, Tokyo (2002)
[JB23] Akama, S.: Introduction to Image Information Processing (Japanese). Syuuwa, Tokyo (2002)
[JB24] Akama, S.: Data Analysis in Excel (Japanese). Muisuri, Tokyo (2002)
[JB25] Akama, S.: Introduction to Cryptographic Programming in Java (Japanese). Syuuwa, Tokyo (2003)
[JB26] Akama, S.: Introduction to System Designs (Japanese). Gihodo, Tokyo (2003)
[JB27] Akama, S.: Introduction to Operating Systems (Japanese). Kogakusha, Tokyo (2003)
[JB28] Akama, S.: Introduction to Algorithms in Java (Japanese). Morikita, Tokyo (2004)s
[JB29] Akama, S.: Introduction to Algorithms in C (Japanese). Kyoritsu, Tokyo (2004)
[JB30] Akama, S., Yamaguchi, Y.: Basic Mathematics with MuPAD (Japanese). Maruzen, Tokyo (2004)
[JB31] Akama, S.: Applied Numerical Computation in Java (Japanese). Gihodo (2004)
[JB32] Akama, S., Ogasawara, M.: Introduction to Simulation with Java and Excel (Japanese). Denkishoin, Tokyo (2005)
[JB33] Akama, S., Tamaki, S., Nagata, Y.: Introduction to Discrete Mathematics (Japanese). Kyoritsu, Tokyo (2006)
[JB34] Akama, S., Yamaguchi, Y.: Introduction to Statistics with R (Japanese). Gihodo, Tokyo (2006)
[JB35] Akama, S.: Foundations of Image Processing (Japanese). Gihodo, Tokyo (2006)
[JB36] Akama, S.: Introduction to Multimedia (Japanese). Kogakusha, Tokyo (2006)
[JB37] Akama, S.: Textbook for Software Engineering (Japanese). Kogakusha, Tokyo (2006)

[JB38] Akama, S.: Image Processing Programming in Java (Japanese). Kogakusha, Tokyo (2007)
[JB39] Akama, S.: Textbook for Java 3D (Japanese). Kogakusha, Tokyo (2007)
[JB40] Akama, S.: Introduction to JAI (Japanese). Kogakusha, Tokyo (2007)
[JB41] Akama, S.: Textbook for VRML (Japanese). Kogakusha, Tokyo (2007)
[JB42] Akama, S.: Introduction to Cryptography with Java (Japanese). Kogakusha, Tokyo (2007)
[JB43] Akama, S.: Introduction to JMF (Japanese). Kogakusha, Tokyo (2007)
[JB44] Akama, S.: Introduction to Object-Orientedness in Java (Japanese). Kogakusha, Tokyo (2007)
[JB45] Akama, S.: Web 3D-CG Language X3D (XML Version) (Japanese). Kogakusha, Tokyo (2007)
[JB46] Akama, S.: Web 3D-CG Language X3D (VRML Version) (Japanese). Kogakusha, Tokyo (2007)
[JB46] Akama, S.: Beginning with SQL (Japanese). Gihodo, Tokyo (2007)
[JB47] Akama, S.: Textbook for Java Sound (Japanese). Kogakusha, Tokyo (2007)
[JB48] Akama, S.: Textbook for Octave (Japanese). Kogakusha, Tokyo (2007)
[JB49] Akama, S.: Textbook for Java Swing (Japanese). Kogakusha, Tokyo (2008)
[JB50] Akama, S., Miyamoto, S.: Logics for Soft Computing (Japanese). Kogakusha, Tokyo (2008)
[JB51] Akama, S.: Introduction to Statistical Analysis with Excel (Japanese). Kogakusha, Tokyo (2008)
[JB52] Akama, S.: Java 2D/3D Programming (Japanese). Kogakusha, Tokyo (2008)
[JB53] Akama, S.: Learning Data Structures and Algorithms by Java (Japanese). Kogakusha, Tokyo (2008)
[JB54] Akama, S.: Introduction to Complex Systems by Java (Japanese). Kogakusha, Tokyo (2008)
[JB55] Akama, S.: Learning AI Programming with Prolog (Japanese). Kogakusha, Tokyo (2008)
[JB55] Akama, S.: Introduction to Simulations by Octave (Japanese). Kogakusha, Tokyo (2008)
[JB56] Akama, S.: A Mannuaal against Internet Criminals (Japanese). Kogakusha, Tokyo (2008)
[JB57] Akama, S.: Basic Knowledge in Computer Age, Revised edn (Japanese). Corona, Tokyo (2009)
[JB58] Akama, S.: Textbook for Databases (Japanese). Kogakusha, Tokyo (2009)
[JB59] Akama, S.: Learning Econometrics by R (Japanese). Kogakusha, Tokyo (2009)
[JB60] Akama, S.: Introductory Course for Scilab (Japanese). DenpaShinbunsha, Tokyo (2009)
[JB61] Akama, S.: Introduction to TCP/IP (Japanese). Kogakusha, Tokyo (2009)
[JB62] Akama, S.: Textbook for Functional Programming (Japanese). Kogakusha, Tokyo (2009)
[JB63] Akama, S.: Introduction to Mathematics for Programmers (Japanese). Kogakusha, Tokyo (2009)

[JB64] Akama, S.: Introduction to Image Processing with Octave (Japanese). Kogakusha, Tokyo (2010)
[JB65] Akama, S.: Functional Programming Language F# (Japanese). Kogakusha, Tokyo (2010)
[JB66] Akama, S.: Introduction to Information Theory (Japanese). Kogakusha, Tokyo (2010)
[JB67] Akama, S.: Numerical Computation System Scilab (Japanese). Kogakusha, Tokyo (2010)
[JB68] Akama, S.: Understanding Quantum Computers (Japanese). Kogakusha, Tokyo (2010)
[JB69] Akama, S.: Learning Computer Algebras by Maxima (Japanese). Kogakusha, Tokyo (2010)
[JB70] Akama, S.: Understanding Ontology (Japanese). Kogakusha, Tokyo (2010)
[JB71] Akama, S.: Knowledge of Programmers (Japanese). Kogakusha, Tokyo (2010)
[JB72] Akama, S.: Introduction to Artificial Life (Japanese). Kogakusha, Tokyo (2010)
[JB73] Akama, S.: Beginning wit C# (Japanese). Kogakusha, Tokyo (2010)
[JB74] Akama, S.: Specification Language Z (Japanese). Kogakusha, Tokyo (2010)
[JB75] Akama, S.: Understanding Data Mining (Japanese). Kogakusha, Tokyo (2010)
[JB76] Akama, S.: Textbook of System Designs (Japanese). Kogakusha, Tokyo (2011)
[JB77] Akama, S.: Introduction to Semantic Webs (Japanese). Cutt System, Tokyo (2011)
[JB78] Akama, S.: Beginning with C# Graphics (Japanese). Kogakusha, Tokyo (2011)
[JB79] Akama, S.: Learning Calculus by Maxima (Japanese). Kogakusha, Tokyo (2011)
[JB80] Akama, S.: Learning Linear Algebras by Maxima (Japanese). Kogakusha, Tokyo (2011)
[JB81] Akama, S.: Easy Introduction to R (Japanese). Cutt System, Tokyo (2011)
[JB82] Akama, S.: Beginning with Programming by Scala (Japanese). Kogakusha, Tokyo (2011)
[JB83] Akama, S.: Introduction to Mechanics by Maxima (Japanese). Kogakusha, Tokyo (2011)
[JB84] Akama, S.: Beginning wit FreeMat (Japanese). Kogakusha, Tokyo (2011)
[JB85] Akama, S.: R Reference Book (Japanese). Cutt System, Tokyo (2011)
[JB86] Akama, S.: Understanding BABOK (Japanese). Kogakusha, Tokyo (2011)
[JB87] Akama, S.: Textbook of Natural Language Processing (Japanese). Kogakusha, Tokyo (2011)
[JB88] Akama, S.: Introduction to Risk Engineering (Japanese). Kogakusha, Tokyo (2011)
[JB89] Akama, S.: Textbook of Artificial Intelligence (Japanese). Kogakusha, Tokyo (2012)

[JB90] Akama, S.: Super Introduction to Computers (Japanese). Kogakusha, Tokyo (2012)
[JB91] Akama, S.: Introduction to Electromagnetism by Maxima (Japanese). Kogakusha, Tokyo (2012)
[JB92] Akama, S.: Free Software Computing (Japanese). Kogakusha, Tokyo (2012)
[JB93] Akama, S.: Beginning with Processing (Japanese). Kogakusha, Tokyo (2012)
[JB94] Akama, S.: Introduction to Formal Methods (Japanese). Kogakusha, Tokyo (2012)
[JB95] Akama, S.: Learning Differential Equations (Japanese). Kogakusha, Tokyo (2013)
[JB96] Akama, S.: It Processing GUI Programming (Japanese). Kogakusha, Tokyo (2013)
[JB97] Akama, S.: Introduction to Multimedia, Extended Version (Japanese). Kogakusha, Tokyo (2013)
[JB98] Akama, S.: Beginning with Statistics by R (Japanese). Kogakusha, Tokyo (2013)
[JB99] Akama, S.: Statistics from the Basics (Japanese). Kogakusha, Tokyo (2013)
[JB100] Akama, S.: Understanding Big Data (Japanese). Kogakusha, Tokyo (2014)
[JB101] Akama, S.: Introduction to Collective Intelligence (Japanese). Kogakusha, Tokyo (2014)
[JB102] Akama, S.: Manual of Writing a Paper by LATEX (Japanese). Kogakusha, Tokyo (2014)
[JB103] Akama, S.: Introduction to Programming by R (Japanese). Kogakusha, Tokyo (2014)
[JB104] Akama, S.: Understanding DNA Computers (Japanese). Kogakusha, Tokyo (2015)
[JB105] Akama, S.: Understanding Wavelet Transformations (Japanese). Kogakusha, Tokyo (2016)
[JB106] Akama, S.: Understanding Wavelet Transformations with Practical Applications (Japanese). Kogakusha, Tokyo (2016)
[JB107] Akama, S.: Introduction to Fintech (Japanese). Kogakusha, Tokyo (2017)
[JB108] Akama, S.: Introduction to Financial Engineering (Japanese). Kogakusha, Tokyo (2018)
[JB109] Akama, S.: Introduction to Rough Sets I (Japanese). Kogakusha, Tokyo (2019)
[JB110] Akama, S.: Introduction to Rough Sets II (Japanese). Kogakusha, Tokyo (2019)

Acknowledgements We are grateful to Prof. Seiki Akama and Dr. Yotaro Nakayama for giving valuable comments.

References

1. Abe, J.M.: On the Foundations of annotated logics (Portuguese). Ph.D. thesis, University of Sao Paulo, Brazil (1992)
2. Akama, S.: Definite description in Montague grammar. B.Sc. thesis, Science University of Tokyo, Japan (1984)
3. Akama:, S.: Constructive predicate logic with strong negation and model theory. Unpublished B.Sc. thesis, Science University of Tokyo, Japan (1984). Later published as [JP2]
4. Akama, S.: Constructive falsity: foundations and their applications to computer science. Ph.D. thesis, Keio University, Japan (1989)
5. da Costa, N.C.A., Subrahmanian, V.S., Vago, C.: The paraconsistent logic $P\mathcal{T}$. Zeitschrift für mathematische Logik und Grundlagen der Mathematik **37**, 139–148 (1991)
6. da Costa, N.C.A., Abe, J.M., Subrahmanian, V.S.: Remarks on annotated logic. Zeitschrift für mathematische Logik und Grundlagen der Mathematik **37**, 561–570 (1991)
7. Dowty, D., Wall, R., Peters, S.: Introduction to Montague Semantics. Reidel, Dordrecht (1980)
8. Hilbert, D., Ackermann, W.: Grundzügen der theoretischen Logik. Springer-Verlag, Brlin (1928)
9. Novikov, P.: Elements of mathematical logic (Russian) (1959). Englisj version was published in 1964 by Addison-Wesley

Chapter 2
A Busy-Beaver-Like Function in Complexity Theory

Francisco A. Doria, Carlos Alberto Nunes Cosenza, and Luis Claudio Bernardo Moura

Contents

Abstract In this work we will briefly describe the counterexample function f for $P = NP$, which is one of those functions of the Busy Beaver type, because it exceeds all the computable functions in its peaks, and it is also a non-computable function.

Keywords Non-computable function, Computability, $P = NP$ problem, Busy Beaver functions

2.1 Introduction

The Busy Beaver function has two important properties:

- It dominates all computable total functions, and is,
- As a consequence, a noncomputable function [11].

We call all functions that satisfy those two conditions *Busy-Beaver-like* functions.

We are going to briefly describe *the counterexample function* f to $P = NP$, which is one of those Busy Beaver-like functions, as it overtakes in its peaks allcomputable

[1] Around 1992.

F. A. Doria (✉) · C. A. N. Cosenza · L. C. B. Moura
LabFuzzy, PEP-COPPE, UFRJ, Rio de Janeiro, Brazil
e-mail: fadoria63@gmail.com

J. M. Abe (ed.), *Advances in Applied Logics*, Intelligent Systems Reference Library 243,
https://doi.org/10.1007/978-3-031-35759-6_2

functions, and is also a noncomputable function. However it arises out of a totally different context as what we can call "the Busy Beaver Game."

Kreisel pointed out to N. da Costa and F. A. Doria in a private exchange by e-mail[1] that this counterexample function f has the fast-growing property, and might be of import in several issues in complexity theory. He gave no proofs for his claims about f—and we sort of informally decided to refer to Kreisel's assertions as *Kreisel's Conjectures*. It took us some time to prove it. Such proof appear in several papers, e.g. [6].

We give those arguments here, as we believe they are of interest per se and might be of interest for possible applications in computer science,

The first, immediate consequence of Kreisel's Conjectures is:

Proposition 2.1.1 *If the counterexample function f is proved total within an axiomatic theory S and if the recursive function g can be embedded into f, then g is proved total in S.* □

(We'll clarify the meaning of "embedding a function" in the remaining sections of this paper.)

A review paper

This is essentially a review paper where we restate a few—let us say, peculiar and curious—results that lead to Kreisel's Conjectures and beyond. We use the references, which are liberally quoted, as the first author is a co-author or single author of the quoted papers. We stress: the main theoretical constructions have already been published; also, as they are due to the first author, we collect the puzzle's pieces into something that makes sense.

About ten years ago N. da Costa and F. A. Doria made explicit the main results from their exchanges with Kreisel and after circulating them they published those ideas in a preprint [5]. It contains the essential from our Kreisel exchanges. We also had the assistance of Marcel Guillaume, who acted as a severe, rigorous but helpful referee.[2]

The formal context

We suppose that our formal environment is a first-order classical theory S that contains Peano Arithmetic and is sound. Both PA—Peano Arithmetic—and ZFC will be enough.

2.2 Required Notation and Concepts: Function F

Function F below first appears in the proof of Kleene's incompleteness theorem [9]; the current presentation was motivated by Kaye [7].

Logical notation follows standard usage. For more specific notational conventions see [4, 10]. Function F is defined below:

---

[2] Of course mistakes are our own responsibility.

Remark 2.2.1 For each n, $\mathsf{F}(n) = \max_{k \le n}(\{e\}(k)) + 1$, that is is the sup of those $\{e\}(k)$ such that:

1. $k \le n$.
2. $\ulcorner \mathrm{Pr}_S(\ulcorner \forall x\, \exists z\, T(e, x, z) \urcorner) \urcorner \le n$.

$\mathrm{Pr}_S(\ulcorner \xi \urcorner)$ translates as, there is a proof of ξ in S, where $\ulcorner \xi \urcorner$ means: the Gödel number of ξ. So $\ulcorner \mathrm{Pr}_S(\ulcorner \xi \urcorner) \urcorner$ means: "the Gödel number of sentence 'there is a proof of ξ in S.'" Condition 2 above translates as: there is a proof of [$\{e\}$ is total] in S whose Gödel number is $\le n$. □

This construction is a more elaborate version of Kleene's example of a function that appears in his version of Gödel's incompleteness theorem [2, p. 287; 8].

We can prove, with reference to axiomatic system S:

- We can explicitly compute a Gödel number e_{F} so that $\{e_{\mathsf{F}}\} = \mathsf{F}$.
- If S is consistent then $S \nvdash \forall m \exists n\, [\{e_{\mathsf{F}}\}(m) = n]$.

Function F is obviously tied to the, let us say, inner mechanisms of S, which may be seen as coded in the structure of F.

Similarities between BB, the Busy Beaver Function, and F

However we do *not* get a Busy Beaver like function here; what we get is a partial recursive function (and the Busy Beaver is noncomputable…) which can neither be proved nor disproved total in S—it is total in the standard model for arithmetic, provided that S has a model with standard arithmetic.

Even so there are similarities between BB and F:

- BB dominates all intuitively total recursive functions.
- However, F—intuitively, naïvely—dominates all *provably* total recursive functions in our axiomatic system S.

How do we get Busy Beaver like functions within a formal theory like S? Is there some kind of general procedure? After all, F is full of undecidable properties, and, we may say, is conceptually very close to the full BB.

Constructions, I

We will now construct a quite peculiar function within our axiomatic theory and analogous functions. We will now go back to the obscurities of a formal language. We use as our starting point the standard formal definition for $P = NP$, which is due to L. Stockmeyer.

We deal here with several possible formalizations for $P = NP$ and $P < NP$; we have called the unusual formalizations (there are infinitely nonequivalent many such) the "exotic formalizations." They are intuitively equivalent, but when we formalize things in a system like S there are difficulties to be dealt with when we try to establish their equivalence.

However there is one unifying property behind all these diverse formalizations of $P = NP$: they all share the same interpretation in the standard model for arithmetic.

A caveat

The reason we actually have such a manifold (not always formally equivalent) definitions has to do with the fact that when we say[3] that we have a polynomial bound we are talking about, we mean *some* polynomial bound, and not, say, a minimal bound. That apparently minor fact is the source of many complications and introduces many nontrivial complications in our arguments.

Constructions, II

Let $t_m(x)$ be the primitive recursive function that gives the operation time of $\{m\}$ over an input x of length $|x|$ (here we give the standard definition):

Recall that the operation time of a Turing machine is given as follows: if $\{m\}$ stops over an input x, then:

$$t_m(x) = |x| + [\text{number of cycles of the machine until it stops}].$$

t_m is primitive recursive and can in fact be defined out of Kleene's T predicate.

The formalized definitions $[P = NP]$ *and* $[P < NP]$

Definition 2.2.2 (*Standard formalization for* $P = NP$)
$[P = NP] \leftrightarrow_{\text{Def}} \exists m, a \in \omega \forall x \in \omega [(t_m(x) \leq |x|^a + a) \wedge R(x, m)]$. □

$R(x, y)$ is a polynomial predicate; as its interpretation we can say that it formalizes a kind of "verifying machine" that checks whether or not x is satisfied by the output of $\{m\}$. (There is an equivalent formalization for $P = NP$ where one uses Kleene's T predicate to get the time measure t_m.)

Definition 2.2.3 $[P < NP] \leftrightarrow_{\text{Def}} \neg[P = NP]$. □

Now suppose that $\{e_{\mathsf{f}}\} = \mathsf{f}$ is total recursive and strictly increasing:

Remark 2.2.4 The naïve version for the exotic formalization is[4]:

$$[P = NP]^{\mathsf{f}}_{\text{Def}} \leftrightarrow \exists m \in \omega, a \forall x \in \omega [(t_m(x) \leq |x|^{\mathsf{f}(a)} + \mathsf{f}(a)) \wedge R(x, m)].$$

Yet there is no reason why we should ask that f be total besides asking it to be recursive; on the contrary, there will be interesting situations where such a function may be partial and yet it may provide a reasonable exotic formalization for $P < NP$ [4]. □

So, for the next definitions and results let f be in general a (possibly partial) recursive function which is strictly increasing over its domain, and let e_{f} be the Gödel number of an algorithm that computes f. Let $p(\langle e_{\mathsf{f}}, b, c\rangle, x_1, x_2, \ldots, x_k)$ be an universal Diophantine polynomial with parameters e_{f}, b, c; that polynomial has integer

[3] See below.

[4] The exotic formalization is due to da Costa and Doria [3].

roots if and only if $\{e_f\}(b) = c$. Also we may if needed suppose that polynomial to be ≥ 0. Finally we omit the "$\in \omega$" in the quantifiers, since they all refer to natural numbers.

We barely need to fill in some kind of verbal explanation, as all constructions here are purely formal.

Definition 2.2.5 $M_{\mathsf{f}}(x, y) \leftrightarrow_{\mathrm{Def}} \exists x_1, \ldots, x_k \, [p(\langle e_{\mathsf{f}}, x, y \rangle, x_1, \ldots, x_k) = 0]$. □

$M_{\mathsf{f}}(x, y)$ stands for $M_{e_{\mathsf{f}}}(x, y)$, or better, $M(e_{\mathsf{f}}, x, y)$, as dependence is on the Gödel number e_{f}.

Definition 2.2.6 $\neg Q(m, a, x) \leftrightarrow_{\mathrm{Def}} [(t_m(x) \leq |x|^a + a) \rightarrow \neg R(x, m)]$. □

Proposition 2.2.7 (standard formalization, again.)

$$[P < NP] \leftrightarrow \forall m, a \, \exists x \, \neg Q(m, a, x).$$ □

Definition 2.2.8 $\neg Q_{\mathsf{f}}(m, a, x) \leftrightarrow_{\mathrm{Def}} \exists a' \, [M_{\mathsf{f}}(a, a') \wedge \neg Q(m, a', x)]$. □

We will sometimes write $\neg Q(m, \mathsf{f}(a), x)$ for $\neg Q_{\mathsf{f}}(m, a, x)$, whenever f is proved total in S.

Definition 2.2.9 (*Exotic formalization.*)

$$[P < NP]^{\mathsf{f}} \leftrightarrow_{\mathrm{Def}} \forall m, a \, \exists x \, \neg Q_{\mathsf{f}}(m, a, x).$$ □

Again this is a Π_2 arithmetic sentence, very much like the standard definition $[P < NP]$:

$$\forall m, a \, \exists x, a', x_1, \ldots, x_k \, \{[p(\langle e_{\mathsf{f}}, a, a' \rangle, \ldots, x_1, \ldots, x_k) = 0] \wedge \neg Q(m, a', x)\}.$$

(Recall that Q is primitive recursive.)

Definition 2.2.10 $[P = NP]^{\mathsf{f}} \leftrightarrow_{\mathrm{Def}} \neg [P < NP]^{\mathsf{f}}$. □

2.3 Kreisel's Counterexample Function to $[P = NP]$

For the definition of SAT see [4, 10]; for the BGS recursive set of poly Turing machines see also [2]. In a nutshell, SAT is the set of all Boolean expressions in conjunctive normal form (cnf) that are satisfiable, and BGS is a recursive set of poly Turing machines that contains emulations of every conceivable poly Turing machines.

The definition of f

The full counterexample function f is defined as follows: let ω be also a set of codes for an enumeration of the Turing machines (on what we mean by "standard coding," [5]). Any standard coding of Turing machines is a monotonic map from the set of machines onto ω. We refer to [5] for details. f is defined as follows:

- If $n \in \omega$ isn't a poly machine, $f(n) = 0$.
- If $n \in \omega$ codes a poly machine:
  - $f(n) =$ first instance x of SAT so that the machine fails to output a satisfying line for x, plus 1, that is, $f(n) = x + 1$.
  - Otherwise $f(n)$ is undefined, that is, if $P = NP$ holds for n, $f(n) =$ undefined.

f is non computable. It will also turn out to be at least as fast growing as the Busy Beaver BB function, since in its peaks it surpasses all intuitively total recursive functions.

2.4 Main Steps in Our Argument

We present our argument to deal with f in an intuitive (or naïve) form: actually it is presented in the form below whenever we deal with those avatars of the BB function—after all, it is only a kind of recipe.

- Use the *s–m–n* theorem to obtain Gödel numbers for an infinite family of "quasi-trivial machines"—soon to be defined. The table for those Turing machines requires very large numbers, and the goal is to get a compact code for that value in each quasi-trivial machine so that their Gödel numbers are in an increasing sequence $\mathsf{g}(0), \mathsf{g}(1), \mathsf{g}(2), \ldots$. Here g is chosen to be primitive recursive.
- Then add the required clocks as in the BGS sequence of poly machines, and get the Gödel numbers for the pairs machine + clock. We can embed the sequence we obtain into the sequence of all Turing machines.
- Notice that the subsets of poly machines we are dealing with are (intuitive) recursive subsets of the set of all Turing machines. More precisely: if we formalize everything in some theory S, then the formalized version of the sentence "the set of Gödel numbers for these quasi-trivial Turing machines is a recursive subset of the set of Gödel numbers for Turing machines" holds of the standard model for arithmetic in S, and vice versa.
  However S may not be able to prove or disprove that assertion, that is to say, it will be formally independent of S.[5]
- We can thus define the counterexample function(s) over the desired set(s) of poly machines, and we get a rather weird collection where each Turing machine is represented by infinite many avatars.

A peculiar property of these BGS machines is: given a Turing machine

Domination

Definition 2.4.1 For $f, g : \omega \to \omega$,

$$f \textbf{ dominates } g \leftrightarrow_{\text{Def}} \exists y \, \forall x \, (x > y \to f(x) \geq g(x)).$$

---

[5] That's is a tricky construction indeed!

We write $f \succ g$ for f dominates g. □

Quasi-trivial machines

We go on with our recipe: recall that the operation time of a Turing machine is given as follows: if M stops over an input x, then the operation time over x,

$$t_{\mathsf{M}} = |x| + \text{number of cycles of the machine until it stops.}$$

Example 2.4.2 Quasi-trivial machines. A *quasi-trivial machine* Q operates as follows: for $x \leq x_0$, x_0 a constant value, $\mathsf{Q} = \mathsf{R}$, R an arbitrary total machine. For $x > x_0$, $\mathsf{Q} = \mathsf{O}$ or O'.

This machine has a linear bound for its operation time. □

Remark 2.4.3 Now let H be any fast-growing, superexponential total machine. Let H' be a total Turing machine. Form the following family $\mathsf{Q}_{\ldots}$ of quasi-trivial Turing machines with subroutines H and H':

1. If $x \leq \mathsf{H}(n)$, $\mathsf{Q}^{\mathsf{H},\mathsf{H}',n}(x) = \mathsf{H}'(x)$;
2. If $x > \mathsf{H}(n)$, $\mathsf{Q}^{\mathsf{H},\mathsf{H}',n}(x) = 0$. □

Proposition 2.4.4 *There is a family* $\mathsf{R}_{\mathsf{g}(n,|\mathsf{H}|,|\mathsf{H}'|)}(x) = \mathsf{Q}^{\mathsf{H},\mathsf{H}',n}(x)$, *where* g *is primitive recursive, and* $|\mathsf{H}|$, $|\mathsf{H}'|$ *denotes the Gödel number of* H *and of* H'. □

2.5 The Crucial Step

We are interested in quasi-trivial machines where $\mathsf{H}' = \mathsf{T}$, the standard truth-table exponential algorithm for SAT. If the counterexample function is defined over all Turing machines (with the extra condition that the counterexample function equals 0 if M_m isn't a poly machine), we have:

Proposition 2.5.1 *If* $\mathsf{g}(n)$ *is the Gödel number of a quasi-trivial machine as in Remark 2.4.3, then* $f(\mathsf{g}(n)) = \mathsf{H}'(n) + 1$.

Proof Use the machines in Proposition 2.4.4 and Remark 2.4.3. □

This is the main tool we use in order to embed total recursive functions within f.

2.6 f Is a Busy-Beaver-Like Function

We now want to prove the following result: *no total recursive function dominates the counterexample function* f. This is *Kreisel's Conjecture for* f. We do so in order to prove that the counterexample function f overreaches all provably total recursive functions as required of a busy-beaver-like function.

Remark 2.6.1 Sketch of proof:
The main idea goes as follows: suppose that there is a total recursive function $\mathsf{h}(n)$ that dominates f. Get a total recursive $\mathsf{k}(n)$ that dominates h and so that the relative growth speed of k with respect to h is faster that any primitive recursive function.

Why do we need such a condition? Why do we need these primitive recursive functions? We use the quasi-trivial machines to reproduce k within f, that is, we (sort of) replicate function $\langle n, \mathsf{k}(n)\rangle$ within f by a sequence of machines with Gödel numbers $N(n), n = 0, 1, 2, \ldots$ (see above Proposition 2.5.1), where N is primitive recursive, so that we have that k becomes the sequence of machines $N(n), n = 0, 1, 2, \ldots$, and we get the value of f at k as $\langle N(n), \mathsf{k}(n) + 1\rangle$ with $f(N(n)) = \mathsf{k}(n) + 1$.

And so we are done!

As N is primitive recursive by construction, and monotonic increasing on n—these are the crucial properties—it slows down the growth of k by a primitive recursive function. Given our construction—which is trivially fulfilled—we have that f still overtakes h infinitely many times, as k grows faster than h, and we are done.

Embed k into f and obtain a subfragment of f which we note f_{f_k}, one immediately notices that f_{k} grows like k but for a primitive recursive function.

2.7 Proof of Kreisel's Conjecture for f

Lets state the conjecture:

Proposition 2.7.1 *For no total recursive function* h *does* $\mathsf{h} \succ f$.

Proof Suppose that there is a total recursive function h such that we can construct the desired functions with the tools described above.

Follows an example and immediately after our main argument.

Example 2.7.2 For given h, we obtain out of that total recursive function by the usual constructions a strictly increasing total recursive h^*. Then if, for instance, F_ω is Ackermann's function, $\mathsf{h}' = \mathsf{h}^* \circ \mathsf{F}_\omega$ will do. (The idea is that F_ω dominates all primitive recursive functions, and therefore h^* composed with it dominates $\mathsf{g}(n)$.) □

We have that the Gödel numbers of the quasi-trivial machines Q are given by $\mathsf{g}(n)$. Choose adequate quasi-trivial machines, so that $f(\mathsf{g}(n)) = \mathsf{h}'(n) + 1$, from Proposition 2.5.1. We now conclude our argument. If we make explicit the computations, for $\mathsf{g}(n)$ (as the argument holds for any strictly increasing primitive recursive g):

$$f(\mathsf{g}(n)) = \mathsf{h}'(n) + 1 = \mathsf{h}^*(\mathsf{F}_\omega(n)) + 1,$$

and

$$\mathsf{h}^*(\mathsf{F}_\omega(n)) > \mathsf{h}^*(\mathsf{g}(n)).$$

For $N = \mathsf{g}(n)$,

$$f(N) > \mathsf{h}^*(N) \geq \mathsf{h}(N), \text{ all } N.$$

Therefore no such h can dominate f. □

We emphasize it:

Proposition 2.7.3 *No total recursive function dominates f.* □

So, f contains a lot of information about the total recursive functions in our system.

BGS-*like sets; extended Kreisel's Conjecture*

We now extend our analysis to include exotic BGS machines.[6]

We keep following [5]. The main extended Kreisel's conjectures are those for the BGS set of poly machines:

$$\langle \mathsf{M}_m, |x|^a + a \rangle,$$

where we couple a Turing machine M_m to a clock regulated by the polynomial $|x|^a + a$, that is, it stops M_m after $|x|^a + a$ steps in the operation over x, where x is the machine's binary input and $|x|$ its bit-length.

To say it in different words: f contains data about all total recursive functions. And if it contains all data about total recursive functions, it contains lots of data about mathematical systems which are mirrored in those functions. And so on. (These ideas are found in the classic, path-breaking Baker, Gill and Solovay paper [1].)

More on alternative formulations

A more general machine-clock couple will now be dealt with here:

$$\langle \mathsf{M}_m, |x|^{(a)} + \mathsf{f}(a) \rangle \mapsto \mathsf{M}_{\mathsf{c}(m, |\mathsf{f}|, a)},$$

Its Gödel number is given by $\mathsf{c}(m, |\mathsf{f}|, a)$, with c primitive recursive by the *s–m–n* theorem.[7]

Remark 2.7.4 Notice that we can have c such that, for parameters a, b, if $a < b$, then $\mathsf{c}(\ldots a \ldots) < \mathsf{c}(\ldots b \ldots)$. □

The formalized $[P < NP]$ sentence is given by a Π_2 arithmetic sentence, that is, a sentence of the form "for every x there is an y so that $P(x, y)$," where $P(x, y)$ is a very simple kind of relation.[8] Now given a theory S with enough arithmetic in it, S proves a Π_2 sentence ξ if and only if the associated Skolem function f_ξ is proved to be total recursive by S.

We notice that for $P < NP$, the Skolem function is what we have been calling the counterexample function.

---

[6] Named after Baker, Gill and Solovay [1].

[7] The argument that follows first appeared in [4].

[8] It is a primitive recursive predicate.

However there are infinitely many counterexample functions we may consider, an *embarras de choix*, as they say in French given each exotic formulation. Why is it so? For many adequate, reasonable theories S, we can build a recursive (computable) *scale of functions*. Such a "scale of functions" exists and can be explicitly constructed out of the definition–construction for F: $\mathsf{f}_0, \mathsf{f}_1, \ldots, \mathsf{f}_k, \ldots$ with an infinite set of S-provably total recursive functions so that f_0 is dominated by f_1 which is then dominated by f_2, ..., and so on.

Actually notice that there is one such "scale of functions" whenever we can construct a function like F.

Step by step procedure

We quote and summarize it from a previous work of da Costa and Doria.

> Given each function f_k, we can form a BGS-like set BGS^k, where clocks in the time-polynomial Turing machines are bounded by a polynomial:
>
> $$|x|^{\mathsf{f}_k(n)} + \mathsf{f}_k(n),$$
>
> and $|x|$ denotes the length of the binary input x to the machine. We can then consider the recursive set:
>
> $$\bigcup_k \text{BGS}^k$$
>
> of all such sets.
>
> Each BGS^k contains clones of all poly machines (time polynomial Turing machines). Now, what happens if:
>
> - There is a function g which is total provably recursive in S and which dominates all segments f_k of counterexample functions over each BGS^k ?
> - There is no such an g, but there are functions g_k which dominate each particular f_k, while the sequence $\mathsf{g}_0, \mathsf{g}_1, \ldots$ is unbounded in S, that is, grows as the sequence $\mathsf{F}_0, \mathsf{F}_1, \ldots$ in S ? (For the sequence as already mentioned see Remark 2.2.1.)

Remarks

- These ideas first appeared in print in [4] and are due to da Costa and Doria. The present construction was developed by da Costa and Doria in order to settle Kreisel's conjectures on the fast-growing behavior of the counterexample function f to $P = NP$ and similar objects. They first appeared in [4]. A more detailed presentation appears in [2]. f has some unexpected properties, that are pondered in our more recent papers. We base our exposition here on that previous work, which is liberally quoted since it was originally developed by N. da Costa and F. A. Doria.
- One must consider alternative possible formal treatments for available formalizations for $P = NP$. Basically we look for formalizations of the intuitive sentence $P = NP$. We discuss one such possibility in [5]. The consequences seem to be different for the several available formalizations.

2.8 More Exoticisms

Again we quote from [5]. We now extend these ideas to the so-called *exotic machines*.

Exotic $\mathrm{BGS}^{\mathsf{F}}$ *machines*

We quote from [5]. Let F be a fast growing, intuitively total, algorithmic function. We consider exotic $\mathrm{BGS}^{\mathsf{F}}$ machines, that is, poly machines coded by the pairs $\langle m, a\rangle$, which code Turing machines M_m with bounds $|x|^{\mathsf{F}(a)} + \mathsf{F}(a)$. Since the bounding clock is also a Turing machine, now coupled to M_m, there is a primitive recursive map c so that:

$$\langle \mathsf{M}_m, |x|^{\mathsf{F}(a)} + \mathsf{F}(a)\rangle \mapsto \mathsf{M}_{\mathsf{c}(m,|\mathsf{F}|,a)},$$

where $\mathsf{M}_{\mathsf{c}(m,|\mathsf{F}|,a)}$ is a explicit poly machine within the recursive listing of all Turing machines.

Proposition 2.8.1 *Given the counterexample function* f_k *defined over the* BGS^k*-machines, for no S-provable total recursive* h *does* $\mathsf{h} \succ \mathsf{f}_k$. □

Remark 2.8.2 Notice that we have a perfectly reasonable formalization for our main question:

$$[P < NP]^k \leftrightarrow [P < NP]^{\mathsf{f}^k}.$$

Also, $S \vdash [P < NP]^k \leftrightarrow [\mathsf{f}_c^k \text{is total}]$. Again our analysis will give estimates for the growth rate of each counterexample function f_c^k. □

Remark 2.8.3 The previous statements have interesting consequences, which we will briefly pursue below. For the proof of the proposition choose a BGS^k so that f_k dominates all strictly increasing fast growing provably total recursive functions that eventually appear in the proof. □

Box of cookies

We now list a few easy miscellaneous results. We can say, for total f_c^k:

Proposition 2.8.4 *For each j there is a k, $k > j + 1$, so that S proves the sentence "f_k doesn't dominate the* BGS^k *counterexample function f_c^k."* □

However we cannot conclude that "for all j, we have that..." since that would imply that S proves "for all j, f_j is total" as a scholium, which cannot be done (as that is equivalent to "F_S is total," which again cannot be proved in S).[9]

What can be concluded: let S' be the theory $S + \mathsf{F}_S$ is total. Then:

Proposition 2.8.5 *If S is consistent and if f_c^k is total in a model with standard arithmetic for each k, then S' proves: there is no proof of the totality of f_c^k, any k, in S.*

[9] Again we have here a delicate point in our discussion.

Proof See below the discussion. □

Remark 2.8.6 Notice that:

- $S' \vdash \forall k\,([P < NP]^k \leftrightarrow [\mathsf{f}_c^k \text{is total}])$, while S cannot prove it.
- $S' \vdash \forall k\,([P < NP]^k \leftrightarrow [P < NP])$ while again S cannot prove it.
- S' is $S + [S$ is Σ_1 sound$]$. □

Remark 2.8.7 That means that we can conclude:

S' proves that, for every k, S cannot prove $[P < NP]^k$.

Now: does the $[P < NP]^k$ adequately translate our main question? □

Remark 2.8.8 Notice that theory $S + \mathsf{F}_S$ is total is $S + S$ is Σ_1-sound. This will have further consequences. □

A few more properties of the counterexample function

f and the infinitely many f_c are very peculiar objects. They are fractal-like in the following sense: the essential data about NP-complete questions is reproduced mirror-like in each of the f (or over each BGS^k). The different BGS^k are distributed over the set of all Turing machines by the primitive recursive function $\mathsf{c}(m, k, a)$.

Also we cannot argue within S that for all k, f_k dominates ..., as that would imply the totality of the recursive function F_S.

Wild explorations

It is interesting to keep in mind a picture of these objects. First notice that the BGS and BGS^k machines are interspersed among the Turing machines. The quasi-trivial Turing machines have their Gödel numbers given by the primitive recursive function $\mathsf{c}(k, n)$—we forget about he other parameters—where:

- k refers to f^k and to BGS^k as already explained;
- n is the argument in $\mathsf{f}^k(n)$.

Therefore fast-growing function f^k is sort of cloned among the values of the BGS^k counterexample function while slightly slowed down by c. (Recall that c is primitive recursive, and cannot compete in growth power with the f^k.)

Function f^k compresses what might be a very large number into a small code given by the Gödel number of g^k and by n (recall that the length of $\mathsf{f}^k(n)$ is the order of $\log \mathsf{f}^k(n)$). The effect is that all functions f^j, $j < k$ embedded into the k-counterexample function via our quasi-trivial machines keep their fast-growing properties and allow us to prove that the counterexample function is fast-growing in its peaks for BGS^k.

For $j > k$ the growth power of f^k doesn't compensate the length of the parameters in the bounding polynomial that regulates the coupled clock in the BGS^k machines.

Finally while $j < k$, the compressed Gödel numbers of the quasi-trivial machines—they depend on the exponent and constant of the polynomial $x^{\mathsf{f}^k(n)} + \mathsf{f}^k(n)$ which regulates the clock—grow much slower that the growth rate of the counterexample function over these quasi-trivial machines (depending on f^j) and so their fast growing properties come out clearly.

2.9 Envoi

It is a pleasure to dedicate the present paper to Professor S. Akama. We thank Professor J. M. Abe for his kind invitation to contribute to this volume. This paper was supported in part by CNPq-MCT/Brazil, Philosophy Section.

References

1. Baker, T., Gill, J., Solovay, R.: Relativizations of the $P =?NP$ question. SIAM J. Comput. **4**, 431–442 (1975)
2. Chaitin, G.J., da Costa, N.C.A., Doria, F.A.: Gödel's Way. CRC Press (2012)
3. da Costa, N.C.A., Doria, F.A.: Consequences of an exotic formulation for $P = NP$. Appl. Math. Comput. **145**, 655–665 (2003)
4. da Costa, N.C.A., Doria, F.A., Bir, E.: On the metamathematics of the P vs. NP question. Appl. Math. Comput. **189**, 1223–1240 (2007)
5. da Costa, N.C.A., Doria, F.A.: Variations on a complex theme, preprint. Production Engineering Program, coppe/UFRJ, Fuzzy Sets Lab (2015)
6. Doria, F.A.: Selva selvaggia. In Wuppuluri, S., Doria, F.A. (eds.) Unravelling Complexity. World Scientific (2020)
7. Kaye, R.: Models of Peano Arithmetic. Clarendon Press (1993)
8. Kleene, S.C.: General recursive functions of natural numbers. Math. Ann. **112**, 727–765 (1936)
9. Kleene, S.C.: Mathematical Logic. Wiley (1967)
10. Machtey, M., Young, P.: An Introduction to the General Theory of Algorithms. North-Holland (1979)
11. Rado, T.: On non-computable functions. Bell Syst. Tech. J. **41**, 877–884 (1962)

Chapter 3
On the Choice of Primitives in Tense Logic

Seiki Akama and Jair Minoro Abe

Contents

Abstract This paper discusses the axiomatization of basic tense logic K_t. If F and P are primitive as McArthur did, then the resulting system is not complete for Kripke semantics. This can be shown as Humberstone pointed out for normal modal logic K. We show correct axiomatization K_t'.

Keywords Basic tense logic · Axiomatization · Kripke semantics

3.1 Introduction

By *basic tense logic*, we mean the tense system K_t originally proposed by Prior [5]. It has four tense operators G, H, F and P. When axiomatizing K_t, there are two options, i.e., one is to take G and H as primitive and the other F and P as primitive. Most people adopted the former option, presented here in Sect. 3.2, but some people (cf. McArthur [4]) the latter one.

A Kripke semantics for K_t can be specified in the obvious way. It is well-known that K_t is complete for Kripke semantics (cf. Gabbay et al. [2]). However, the result may be subtle, since completeness holds for the former but does not for the latter. The issue has been considered in Humberstone [3] for normal modal (tense) logics.

In this paper, we propose correct axiomatization K_t' for basic tense logic in which F and P are primitive. The structure of the paper is as follows. In Sect. 3.2, we give a quick review of tense logic K_t. In Sect. 3.3, we present an alternative axiomatization for it.

S. Akama
C-Republic, Japan
e-mail: akama@jcom.home.ne.jp

J. M. Abe (✉)
Graduate Program in Production Engineering, Paulista University, São Paulo, Brazil
e-mail: jairabe@uol.com.br

J. M. Abe (ed.), *Advances in Applied Logics*, Intelligent Systems Reference Library 243,
https://doi.org/10.1007/978-3-031-35759-6_3

3.2 Basic Tense Logic

Prior's basic tense logic was originally given by an *axiomatic system* (Hilbert system), which includes axioms and rules of inference (cf. Gabbay et al. [2]; Akama et al. [1]).

The language of K_t includes: propositional connectives $\wedge, \vee, \rightarrow, \neg$ and primitive tense operators G and H with F and P defined by $\mathrm{F}A =_{\mathrm{def}} \neg\mathrm{G}\neg A$ and $\mathrm{P}A =_{\mathrm{def}} \neg H \neg A$.

Axiomatic System for K_t
Axioms
(A1) Axioms of classical propositional logic
(A2) $\mathrm{G}(A \rightarrow B) \rightarrow (\mathrm{G}A \rightarrow \mathrm{G}B)$
(A3) $\mathrm{H}(A \rightarrow B) \rightarrow (\mathrm{H}A \rightarrow \mathrm{H}B)$
(A4) $A \rightarrow \mathrm{HF}A$
(A5) $A \rightarrow \mathrm{GP}A$
Rules of Inference
(MP) $\vdash A, \ \vdash A \rightarrow B \ \Rightarrow \vdash B$
(GNEC) $\vdash A \ \Rightarrow \vdash \mathrm{G}A$
(HNEC) $\vdash A \ \Rightarrow \vdash \mathrm{H}A$

Here, $\vdash A$ reads "A is provable in K_t". If the context is ambiguous, we may use a subscript. (MP) is *modus ponens*, and (GNEC) and (HNEC) are *necessitation* for G and H, respectively. The notion of a proof is defined as usual.

Next, we give a Kripke semantics for K_t, although it plays no role in our argument presented in Sect. 3.3. A *Kripke model* for K_t is defined as a triple $M = (T, <, V)$ where T is a set of time points, $<$ is a binary relation on T ($< \subseteq T \times T$), and V is a valuation function: $PV \times T \rightarrow \{1, 0\}$. Here, PV is the set of propositional variables. $V(p, t) = 1$ says that $p \in PV$ is true at $t \in T$ and $V(p, t) = 0$ says that $p \in PV$ is false at $t \in T$.

Valuation can be extended for any formula. We use the notation $M, t \models A$ to means that A is true at t in M. Truth relations of propositional connectives are as usual. Tensed formulas are interpreted as follows:

$$M, t \models \mathrm{G}A \Leftrightarrow \forall u(t < u \Rightarrow M, u \models A)$$
$$M, t \models \mathrm{F}A \Leftrightarrow \exists u(t < u \text{ and } M, u \models A)$$
$$M, t \models \mathrm{H}A \Leftrightarrow \forall u(s < t \Rightarrow M, s \models A)$$
$$M, t \models \mathrm{P}A \Leftrightarrow \exists u(s < t \text{ and } M, s \models A)$$

We say that A is true in M iff $M, t \models A$ for any $t \in T$. We also say that A is valid, written $\models A$, iff it is true in all Kripke models. K_t is shown to be complete, i.e., $\vdash A$ iff $\models A$; see Gabbay et al. [2].

3.3 Alternative Axiomatization

Now, we present K_t by taking F and P as primitive, with G and H defined by: $\mathrm{G}A \leftrightarrow \neg\mathrm{F}\neg A$ and $\mathrm{H}A \leftrightarrow \neg\mathrm{P}\neg A$, where $A \leftrightarrow B$ is shorthand for $(A \rightarrow B) \wedge (B \rightarrow A)$. Such an axiomatization may be found in McArthur [4]. Unfortunately, the corresponding axiomatization in Sect. 3.2 is not complete, as pointed out by Humberstone [3] for normal modal logic K. In fact, $\mathrm{F}A \rightarrow \mathrm{F}\neg\neg A$ and $\mathrm{P}A \rightarrow \mathrm{P}\neg\neg A$ are not provable in the axiomatization. However, Humberstone did not discuss complete axiomatizations.

Our task is to give an alternative axiomatization of K_t by taking F and P as primitive. We call the corresponding system K_t' which requires two additional axioms:

(A6) $\mathrm{F}A \leftrightarrow \mathrm{F}\neg\neg A$
(A7) $\mathrm{P}A \leftrightarrow \mathrm{P}\neg\neg A$.

To show completeness of K_T', it suffices to prove the following theorem.

Theorem 3.1 *In K_t', we can obtain the following:*

(1) $\vdash_{K_t'} \mathrm{F}A \leftrightarrow \neg\mathrm{G}\neg A$
(2) $\vdash_{K_t'} \mathrm{P}A \leftrightarrow \neg\mathrm{H}\neg A$

Proof We only prove (1) and the proof of (2) is similar. We start with a special case of biconditional $\neg\mathrm{F}\neg A \leftrightarrow \mathrm{G}A$, now writing it in the reverse order and taking $\neg A$ in place of A:

$$\neg\mathrm{F}\neg\neg A \leftrightarrow \mathrm{G}\neg A$$

and so negating both sides:

$$\neg\neg\mathrm{F}\neg\neg A \leftrightarrow \neg\mathrm{G}\neg A$$

and cancelling the outer double negation on the left (as we may by classical propositional logic, since it is not the scope of F or G)

$$\mathrm{F}\neg\neg A \leftrightarrow \neg\mathrm{G}\neg A$$

and then again with the newly suggested axiom $\mathrm{F}A \leftrightarrow \mathrm{F}\neg\neg A$. We finally get the desired:

$$\mathrm{F}A \leftrightarrow \neg\mathrm{G}\neg A.$$

As a consequence of theorem 1, completeness of K_t' follows:

Theorem 3.2 *K_t' is complete.*

Theorem 3.2 is a completeness theorem of basic tense logic with P and F as primitive. This can fill the gap of existing incorrect axiomatizations of basic tense logic in the literature. Our proposed tense system K_t' requires two additional axioms with biconditional, not just the one way conditional. It is not economical, but it may be of use for some philosophical applications, e.g., future contingency.

Acknowledgements We are grateful to Prof. Lloyd Humberstone for his valuable comments. We learned the problem explored in this paper from him in 2015, when prepairing the final version of Akama et al. [1].

References

1. Akama, S., Murai, T., Kudo, Y.: Partial and paraconsistent approaches to future contingents in tense logic. Synthese **193**, 3639–3649 (2015)
2. Gabbay, D., Hodkinson, I., Reynolds, M.: Temporal Logic: Mathematical and Computational Aspects, vol. 1. Oxford University Press, Oxford (1994)
3. Humberstone, L.: Yet another "choice of primitives" warning: normal modal logics. Logique et Analyse **185–188**, 395–407 (2004)
4. McArthur, R.: Tense Logic. Reidel, Dordrecht (1976)
5. Prior, A.N.: Past, Present and Future. Clarendon Press, Oxford (1967)

Chapter 4
Paraconsistent Annotated Logic and Chaos Theory: Introducing the Fundamental Equations

João Inácio Da Silva Filho, Mauricio Conceição Mario, Dorotéa Vilanova Garcia, Raphael Adamelk Oliveira, Maurício Fontoura Blos, Hyghor Miranda Côrtes, and Jair Minoro Abe

Contents

Abstract Regarding the Physical Science and Engineering areas, there has been significant interest in finding new ways of applications of Chaos Theory in modeling dynamic systems. In this specific area of knowledge, Paraconsistent Logic (PL) in its expanded form, in which it uses annotations composed of two values (PAL2v), can offer a significant contribution. PAL2v has been applied at the interpretation and modeling of physical systems showing very promising results. In this initial research, we are making analogies between the main foundations of PAL2v and the equation of Logistic Map of Verhulst who originated the modern Chaos Theory and its variations. The results of this comparison show that the procedures used in the application of the

J. I. Da Silva Filho (✉) · M. C. Mario · D. V. Garcia · R. A. Oliveira · M. F. Blos · H. M. Côrtes
Laboratory of Applied Paraconsistent Logic, Santa Cecília University—Unisanta, Oswaldo Cruz Street, Santos, SP 28811045-000, Brazil
e-mail: inacio@unisanta.br

M. C. Mario
e-mail: cmario@unisanta.br

D. V. Garcia
e-mail: dora@unisanta.br

J. M. Abe
Graduate Program in Production Engineering, Paulista University, São Paulo, Brazil

J. M. Abe (ed.), *Advances in Applied Logics*, Intelligent Systems Reference Library 243,
https://doi.org/10.1007/978-3-031-35759-6_4

Logistic Map identify the Chaos Theory with paraconsistent analysis based on the fundamental concepts of PAL2v. Moreover, based on joint interpretations of PAL2v and Chaos Theory, fundamental equations are created called "ParaCaos equations". With the equations obtained by the method of PAL2v analysis in Chaos Theory, there arises significant ways of research about the behavior and stability of chaotic systems. It is observed that the values that identify the PL state of the attractor of stability located on the PAL2v Lattice have a strong relationship with the golden ratio of geometry. Therefore, the equations extracted from the junction PAL2v and Chaos Theory can be used to identify the parameters of stability levels in complex systems and other applications.

Keywords Paraconsistent logic · Paraconsistent annotated logic · Chaos theory · Golden ratio

4.1 Introduction

The basis of our current technology supported by classical science, with its reductionist principles, works with little effectiveness in environments considered complex, which are represented by incomplete information and often contradictory [1–3]. It is known that the proper treatment of complex phenomena has limitations that prevent a complete modeling of these systems, especially with techniques based on classical logic, which has binary characteristics [4, 5].

The efficient models of complex systems, capable of expressing all the richness from the interactions made by information signals, sometimes inconsistent, ask for other forms of interpretation. One of the alternative ways usually used to model systems of this type is known as Complexity Theory or the Chaos Theory [6–10]. Just as in Chaos Theory, the logics that have at their foundations some violation of the strict principles of binary Classical logic are called non-Classical.

The non-classical logics have recently emerged with good results in applications and modeling of complex systems [11]. With these new forms of logical interpretations that depict the real world in a different way, the non-classical logics are used at modeling of the information signals in fields where the classical theory is unable to be applied by a variety of reasons. Among these reasons, can be cited its binary condition that is controlled by laws that do not consider the incompleteness and contradiction [11–13].

Considering the growing interest in practical applications, we can think about developments in computational systems structured in algorithms capable of linking the fundamentals of chaos theory with concepts based solely on a non-classical logic, for instance, the Paraconsistent Logic (PL) [11]. In this context, this work presents the chaos theory joint with a special type of paraconsistent logic, denominated Paraconsistent Annotated Logic with annotation of two values (PAL2v).

PAL2v is seeking ways to develop interpretations of complex systems based on logic that is different from the classical. Therefore, the purpose of this joint is to obtain mathematical models that can be rational and efficient for representing the physical real-world systems [11–13].

This chapter is organized as follows: Sect. 5.2 presents a literature review on Chaos Theory, with the fundamental concepts and main equations, highlighting the equation of Logistic Map and with an application example. In the same section, it is presented the main features of PAL, highlighting the equations of its expanded form, known as PAL2v [11]. Section 3 presents the methods that allow the junction of Chaos Theory with PAL2v. Finally, the simulations results and the conclusions are presented.

4.1.1 Literature Review

For better understanding, this section presents a literature review on the Chaos Theory, the PL and its extensions.

4.1.1.1 Chaos Theory

The study of Chaos Theory began in Mathematics and Physics and was later extended to other areas like Engineering, Computer Science, Social Science, Biology and Medicine [4–6].

Currently there are several application works, as an example, we can mention the use of Chaos Theory in information processing, financial analysis and prediction, as well as control of cardiac activity [6]. It has also been highlighted recent researches based on the Chaos Theory related to the dynamics of control and automation systems applied to the field of engineering, physics and mathematics [6–9].

These studies tried to find ways of modeling the dynamic systems even when they are at levels which may be considered chaotic. So, despite being a new theory, there has been in recent years a significant interest in finding applications of its foundations in systems with dynamic models [6, 10, 14].

The first foundations of Chaos Theory began with the Malthusian model [15], which constituted a mathematical model proposed in 1798 by Briton Thomas Robert Malthus (1766–1834). This proposal, which became one of the attempts to predict the future of an animal population, was, at that time, the cause of some controversy in scientific circles. This first paper was about a linear model where the total population depended on solely on birth rates (A), mortality (B) and number of individuals (N). The mathematical expression for this model of Malthus, which is a geometric progression, is:

$$\frac{dN}{dt} = (A - B)N \tag{5.1}$$

Soon after, in 1845, it was proposed by the Belgian mathematician Pierre François *Verhulst* (1804–1849) a nonlinear model where mortality would be proportional to the square of the number of individuals [16]. *Verhulst*'s model is expressed by the differential equation:

$$\frac{dN}{dt} = AN - BN^2 \tag{5.2}$$

where: N is the number of individuals; A is the birthrate and B mortality rate.

In 1976, the study model of *Verhulst* was taken up with innovations proposed by Robert May [17]. Robert May's new approach has not studied the model in its differential form, but in map form, where each value is obtained from the previous value, so:

$$N_1 = AN_0 - BN_0^2$$

$$N_2 = AN_1 - BN_1^2$$

$$N_{n+1} = AN_n - BN_n^2 \tag{5.3}$$

In this new conception of May, the largest positive value of N will be at the limit of iterations n where the population will be extinct. Therefore, the equation can be written as: $AN_{nmax} - BN_{nmax}^2 = 0$ $AN_{nmax} - BN_{nmax}^2 = 0$.

Since the number of iterations can be related to the time count t, then the above formula can be written considering the maximum time where the population will be extinguished: $AN_{tmax} - BN_{tmax}^2 = 0$.

This condition of largest positive value N means: $N_{tmax} = 0$ or else: $N_{tmax} = \frac{A}{B}$.

Dividing the Eq. (5.3) by this value of N maximum, we have:

$$\frac{N_{t+1}}{N_{tmax}} = A\frac{N_t}{N_{tmax}} - B\frac{N_t^2}{N_{tmax}}\frac{N_{tmax}}{N_{tmax}}$$

t is called x, the ratio: $x_t = \frac{N_t}{N_{tmax}}$.

We have: $x_{t+1} = Ax_t - Bx_t^2\frac{A}{B}$.

Doing: $A = K$.

It is obtained the expression of the Logistic Map:

$$x_{t+1} = Kx_t(1 - x_t) \tag{5.4}$$

It is verified that the Eq. (5.4) is deterministic where x values represent percentage of the population over time, and its future situation will be determined by the conditions of the present.

The K-factor becomes a control parameter where, depending on its value, the behavior of this system can be periodic or chaotic.

As for nonlinear models, except in some special cases, it is not known in an arbitrary time t, or any K, a general solution to the Logistic Map of the Eq. (5.4). However, through successive iterations it is possible to observe the transition to chaotic behavior by tracing the set of attractors of the Logistic Map for different values of K. In the case, one should calculate successive iterations from an initial condition x_0, for x_t:

$$x_1 = f(x_0) = Kx_0(1 - x_0)$$

$$x_2 = f(x_1) = f(f(x_0)) = f^{[2]}(x_0) = Kx_1(1 - x_1)$$

$$x_3 = f(x_2) = f(f(f(x_0))) = f^{[3]}(x_0) = Kx_2(1 - x_2)$$

$$x_t = f(x_t) = f(f(f(...x_0...))) = f^{[t]}(x_0) = Kx_{t-1}(1 - x_{t-1})$$

By applying Eq. (5.4) of the Logistic Map it is verified that for certain values of K, the future behavior of this map loses the regularity and becomes highly sensitive to initial conditions. It is also seen that it is easy to predict future conditions when the behavior of the Logistic Map is periodic, because in this state the values of x obey certain regularity that, in the long term, stabilizes as an attractor. Moreover, in the chaotic regime, any variation in these conditions (initial conditions) generates large variations in future conditions [18]. This makes it very difficult to accurately define the initial conditions. This behavior compromises the predictability of the system which, although deterministic, becomes unpredictable. As this chaotic regime, the attractor loses any regularity and is called strange attractor. Figure 5.1 shows the behavior of x for innumerous t iterations and for various values of K [19].

This transition from periodic regime to chaos is known as the doubling period route, and the point where there is loss of stability of the attractor is called a bifurcation. Therefore, for a study where variations of K are made, the duplications occur in the bifurcation points.

4.1.1.2 Main Analyzes Based on the Concepts of Chaos Theory

The concept of deterministic chaos can be understood as the state of a system when it accuses the existence of erratic behavior or aperiodic and sensitive dependence on initial conditions [14, 19]. In this chaotic state, two processes caused by slightly different initial conditions vary exponentially with t time. This means that when a deterministic model exhibits chaos, problems arise relating to the predictability of future behavior of the system. It is known that in this chaotic state the difference between two initial conditions "slightly different" can be within the range of uncertainty with which these initial conditions are known. In this case, it is not possible to say with certainty which of the two initial conditions is generated the current state.

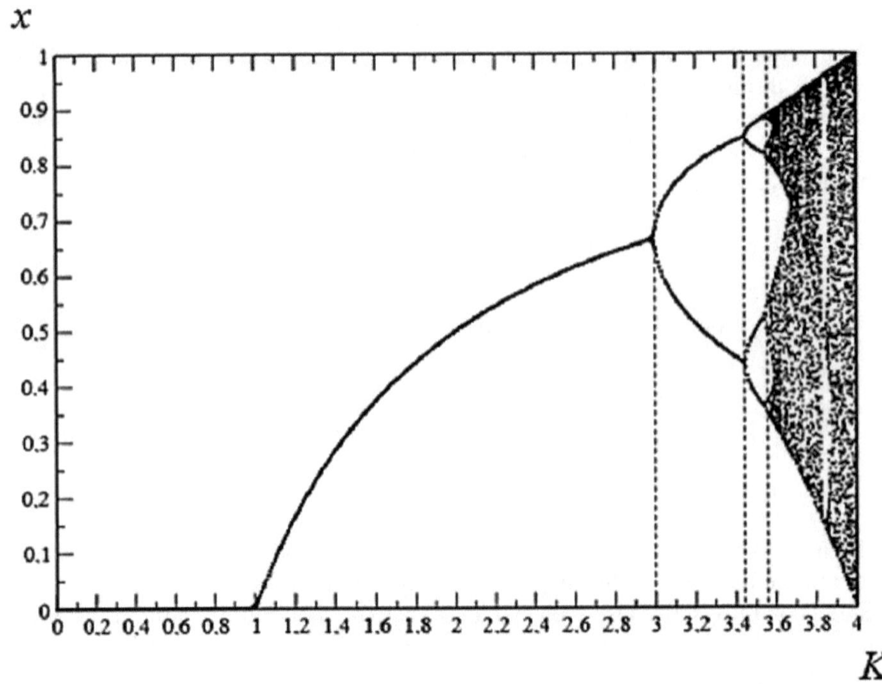

Fig. 5.1 Behavior graph of the output x for numerous t iterations and different values of K

Thus, it is impossible to predict the future behavior of the system, at least for a time arbitrarily long, from these initial conditions.

The Logistic Map of the Eq. (5.4) may be arranged in the form:

$$x_t = f(x_{t-1}) = Kx_{t-1}(1 - x_{t-1})$$

$$x_t = Kx_{t-1}(1 - x_{t-1}) \tag{5.5}$$

where: $0 \leq x_t \leq 1$ and $0 > K \leq 4$.

Equation (5.5) is a second degree polynomial, such that:

$$x_{t+1} = K\left(x - x^2\right) \rightarrow x_{t+1} = Kx - Kx^2$$

When expressed in graph the results of the Eq. (5.5) shows a parabola whose vertex (maximum point) has coordinates: $x_t = \frac{-a}{2a} = \frac{-(-1K)}{2(1K)} = \frac{1}{2}$ and

$$x_{t+1} = -\frac{\Delta}{4a} = \frac{-(-1K)^2 - 4(1 \times 0)}{4(1K)} = \frac{K}{4}$$

Therefore, the maximum point of the parabola of the Logistic Map expressed in Eq. (5.5) is:$(x_{t-1}, x_t) = \left(\frac{1}{2}, \frac{K}{4}\right)(x_{t-1}, x_t) = \left(\frac{1}{2}, \frac{K}{4}\right)$.

By the coordinates it is verified that when $K = 4$ the vertex of the logistic parabola is at $x = 1$, so if $K > 4$ the condition of x being in the range [0,1] is no longer satisfied. Similarly, if $K = 0$ it reduces the parabola to horizontal axis. In this case, for negative values of the K the concavity of the parabola is inverted, which also leads x to values outside the domain [0,1].

4.1.1.3 Fixed Point

A fixed point, given by x^* of a map $x_t = f(x_{t-1})$ is a point that maps itself, ie, such that $x^* = f(x^*)$. If there are points that, when considered in the initial condition of a map $f(x)$ lead, after a finite number of iterations, to the fixed point x^*, these are called fixed final points. Therefore, x is a fixed final point of $f(x)$ if there is a t positive integer such that $f^{[t]}(x)$ is a fixed point$f$ [13, 19, 20]. The fixed point is the solution of the equation: $x^* = f(x^*)$.

As the fixed points represent stationary solutions of a map, whatever it is, then returning to the Logistic Map Eq. (5.5) now represented by fixed point it will be:

$$x_t^* = Kx_{t-1}^*\left(1 - x_{t-1}^*\right)$$

It is verified that the fixed points of the Logistic Map are given by:

1. $x_a^* = 0$, as: $x_a^* = K \times 0_{t-1}(1 - 0_{t-1}) = 0$
2. 2. $x_b^* = 1 - \frac{1}{K}$ with, $K \neq 0$, for:

$$x_b^* = \left(K\left(1 - \frac{1}{K}\right)_{t-1}\right)\left(1 - \left(1 - \frac{1}{K}\right)_{t-1}\right)$$

$$x_b^* = (K - 1)_{t-1}\left(1 - \left(1 - \frac{1}{K}\right)_{t-1}\right)$$

$$x_b^* = \left((K - 1)_{t-1} - \left((K - 1)_{t-1} + \frac{(K - 1)}{K}\right)_{t-1}\right)$$

$$x_b^* = \left(\frac{(K - 1)}{K}\right) = \left(1 - \frac{1}{K}\right)_{t-1}$$

It can be deduced that:

If $K < 1$ follows that it is necessarily negative, in other words, is outside the range of the map definition [0,1]. In this case, only the point fixed at the origin $x_a^* = 0$, exists.

If $1 < K < 4$ there will also be the second fixed point x_b^*.

The analysis of the stability of a fixed point in a system it is important to indicate [19, 20]:

(1) Whether or not it is asymptotically stable, otherwise it will not be achieved by typical initial conditions;
(2) If the iterations of the map converge or not to the fixed point (stable) in a damped or oscillatory form.

Therefore, as seen, the Logistic Map of Eq. (5.5) has two fixed points.

$$x_a^* = 0 \text{ and } x_b^* = 1 - \frac{1}{K} \text{ with } K \neq 0.$$

The derivative with respect to x from the logistic map calculated at the fixed point x^* can provide stability to the origin. For the first fixed point the derivative of the logistic map from Eq. (5.5) is:

$$\frac{df}{dx}\Big|_{x=x_a^*} = K(1-2x)\Big|_{x=x_a^*} = K^* \text{for } 0 < K < 1.$$

How was imposed that $K > 0$, then the fixed point $x = 0$ is stable, since $K < 1$, otherwise it is unstable. When $K = 1$ the linear criterion adopted is not sufficient to determine stability.

4.1.2 Non-Classical Paraconsistent Logic (PL)

Among the many ideas under the Non-Classical Logics it is been created a family of logics that has as its main base the repeal of the principle of non-contradiction, which received the name of Paraconsistent Logic (PL) [11–13]. The initial PL systems containing all logic levels, involving propositional calculations, of predicate and descriptions as well as higher-order logic, due to N.C.A. Da Costa (1954 onwards [11–13, 21].

According to [5, 8] PL is defined as any deductive theory **T** is based on a given logic *L*, and we suppose that all logics considered here contain a connective for negation, symbolized. If two formulas of the language of **T**, one of which is the negation of the other, are both theorems of **T** (*i.e.* for some formula **A**, both **A** and **A** are theorems of **T**), then **T** is said to be *inconsistent*, otherwise **T** is consistent. If all formulas of the language of **T** (or all closed formulas) are theorems of **T**, then **T** is called *trivial*; otherwise it is said to be *non-trivial*. A logic **L** is paraconsistent if it can be the underlying logic of inconsistent, but non-trivial theories. If the theory **T** is inconsistent and non-trivial, then **T** is called a paraconsistent theory [13, 21]. The usual systems of logic, for instance classical and intuitionistic logics, are not paraconsistent.

4.1.2.1 Paraconsistent Annotated Logic (PAL)

The Paraconsistent Annotated Logic (PAL) belongs to the family of PLs and can be represented in a particular way, through a lattice of four vertices in which, intuitively, the constants annotation represented in its vertices will give connotations of extreme Logical states to propositions [11–13, 21].

4.1.2.2 Paraconsistent Annotated Logic with Annotation of Two Values (PAL2v)

As seen in [11, 13, 21] it is possible to obtain through the PAL a representation about how much the notes, or evidences, express knowledge about a proposition *P*. This is accomplished using a lattice formed by ordered pairs of values (μ, λ), which comprise the annotation, as seen in Fig. 5.2.

In this representation, it is fixed an operator ~ : |τ| → | τ | where:

$$\tau = \{(\mu, \lambda) \mid \mu, \lambda \in [0, 1] \subset \Re\}$$

If *P* is a basic formula, then: ~ [(μ, λ)] = (λ, μ) where, $\mu, \lambda \in [0, 1] \subset \Re\}$.
Where ~ has the meaning of negation in PAL.

We introduce the extreme logical Paraconsistent states which are the four vertices of the lattice with Favorable Degree of evidence μ and Unfavorable Degree of evidence λ. We read them in the following way [11]:

- $P_T = P_{(1,1)}$ → The annotation (μ, λ) = (1, 1) assigns intuitive reading that *P* is inconsistent.
- $P_t = P_{(1,0)}$ → The annotation (μ, λ) = (1, 0) assigns intuitive reading that *P* is true.

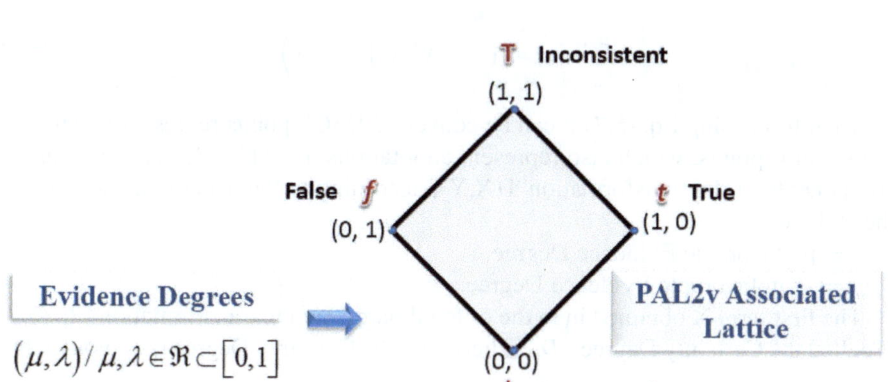

Fig. 5.2 Lattice of four vertexes

- $P_F = P_{(0,\,1)} \rightarrow$ The annotation $(\mu, \lambda) = (0, 1)$ assigns intuitive reading that P is false.
- $P_{\perp} = P_{(0,\,0)} \rightarrow$ The annotation $(\mu, \lambda) = (0, 0)$ assigns intuitive reading that P is Indeterminate.

In the internal point of the lattice, which is equidistant from all four vertices, we have the following interpretation:

- $P_I = P_{(0.5,\,0.5)} \rightarrow$ The annotation $(\mu, \lambda) = (0.5, 0.5)$ assigns intuitive reading that P is undefined.

4.1.2.3 The Logical Negation of the PAL2v

Be $P_{(\mu,\,\lambda)}$ a PL Signal [11–13, 21] where the annotation, composed by the favorable Evidence Degrees μ and unfavorable Evidence Degrees λ, assigns a logical connotation Proposition P, then the logical negation of P is [11, 13]:

$$\neg P_{(\mu,\,\lambda)} = P_{(\lambda,\,\mu)} \tag{5.6}$$

4.1.2.4 The Lattice of Values PAL2v

It is called *Unit Square in Cartesian Plan* (USCP) the Lattice τ with the coordinate system where the values of the favorable Evidence Degree μ are represented on the x axis, and the values of the unfavorable Evidence Degree λ on the y axis (Fig. 5.3). In [11, 13] it was seen that in the paraconsistent system certain annotation (μ, λ) can be identified with the point of the plane in another system κ of values in which the equations can be obtained by a linear transformation, such that:

$$\mathrm{T(X,Y)} = \left(x - y,\ x + y - 1\right) \tag{5.7}$$

Therefore, using Eq. (5.7) it can be converted USCP points representing annotations τ in κ points, which also represent annotations in τ [11–13, 21]. Relating the components of the transformation T(X,Y), according to the usual nomenclature of the PAL2v:

$x = \mu$: favorable Evidence Degree.

$y = \lambda$: unfavorable Evidence Degree.

The first term X obtained in in the ordered pair from the transformation Eq. (5.7) is called the Certainty Degree - D_C. Therefore, the Certainty Degree is computed by:

$$Dc = \mu - \lambda \tag{5.8}$$

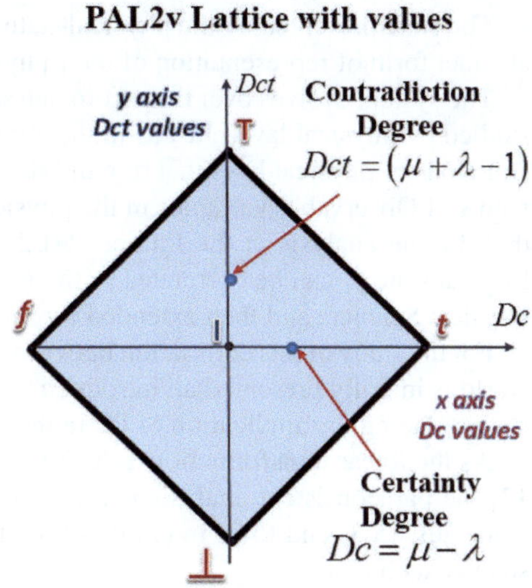

Fig. 5.3 Representation of the axis of the Certainty Degrees (horizontal) and the Contradiction Degrees (Vertical) in the Lattice of κ values

Their values, which belong to the set $\Re$, vary in the range -1 to + 1 and are on the horizontal axis of the lattice of the values, called *"Axis of certainty degrees"*.

The second term Y obtained in the ordered pair of the equation of transformation (7) is called the Contradiction Degree - *Dct*. Therefore, the Contradiction Degree is obtained by:

$$Dct = \mu + \lambda - 1 \tag{5.9}$$

The resulting values of *Dct* belong to the set $\Re$, vary in the closed range + 1 and -1 and are exposed on the vertical axis of the Lattice of values called *"Axis of contradiction degrees."* In the Lattice of values κ when D_C results in + 1 it means that the logical state of paraconsistent analysis is *True*, and when D_C results in -1 means that the logical state of paraconsistent analysis is *False* [11–13, 21].

Similarly, in the Lattice of values κ when *Dct* result in + 1 means that the logical state of the resulting of the paraconsistent analysis is Inconsistent T, and when Dct result in -1 means that the logical state of the resulting of the paraconsistent analysis is Undetermined ⊥ [11–13, 21].

4.1.2.5 The Paraconsistent Logical States ε_τ

The concept of the Paraconsistent Logical state ε_τ is constructed from the fact that in the context of Physical Science, each system or component of a system, holds, or has at any given instant of time, a state [11, 12].

The state presented by the physical status i.e. for a physical body or particle is an abstract form of representation of their physical properties in function of time.

The system evolves over time from one state to another state, and this evolution is studied by physical laws. Just as in the application of PAL2v the information to the equations of physical laws that rule the state of the system are obtained by measurements of Observable Variables in the physical world [21]. It is considered therefore that, by the analysis in the Lattice PAL2v [11, 13] the concept of Paraconsistent Logical state ε_τ can be correlated to the fundamental concept of state, as studied in Physical Science, and then extended for the model based on PL.

For the study of this correlation between the physical world and the Paraconsistent world is initially presented an introduction to the concept of Paraconsistent Logical state ε_τ facing the application of PL in the analysis of physical systems.

As the linear transformation T(X, Y) shown in (7) in the notation of PAL2v [11, 13] the paraconsistent analysis is a function of Evidence Degrees μ and λ. Then, from Eqs. (5.8) and (5.9) in (7) it is possible to represent a logical Paraconsistent state ε_τ such that:

$$\varepsilon_{\tau(\mu,\lambda)} = (\mu - \lambda, \mu + \lambda - 1) \tag{5.10}$$

or else

$$\varepsilon_{\tau(\mu,\lambda)} = (Dc, Dct) \tag{5.11}$$

where: ε_τ is the Paraconsistent Logical state.

Certainty Degree (*D*c) is obtained according to the two Evidence degrees μ and λ.

Contradiction Degree (*D*ct) is found according to the Evidence degrees μ and λ.

In each of the static measurement of Observable Variables in the physical world, in which it is possible to obtain the values of favorable Evidence Degrees μ and unfavorable λ to determine Certainty Degree - D_C and Contradiction Degree - D_{ct}, it is always found a single Paraconsistent Logical state ε_τ related to the two information signals [11–13, 21].

Figure 5.4 presents the paraconsistent Logical state (ε_τ) obtained from two measurements in the physical world.

4.2 The Logistic Map Equation and the Foundations of PAL2v

The analogy to be made between the Logistic Map Eq. (5.5) and the foundations of PAL2v starts with the source of information in which the x values represent percentages of the population. Since the Eq. (5.5) of the Logistic Map is:

$x_t = Kx_{t-1}(1 - x_{t-1})$, then it can rewrite as: $x_t = K\left(x_{t-1} - x_{t-1}^2\right)$.

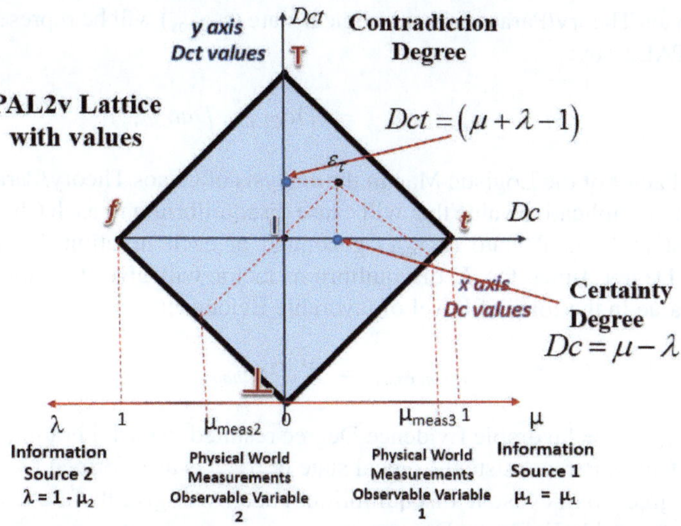

Fig. 5.4 Paraconsistent Logical state ε_τ obtained from two measurements in the physical world

Thus, it is verified that for the analysis with PAL2v, we have:
$\mu_{Chaos} = x_{t-1} \rightarrow$ favorable Evidence Degree (μ).
$\lambda_{Chaos} = x_{t-1}^2 \rightarrow$ unfavorable Evidence Degree (λ).

4.2.1 The ParaChaos Equations

With these considerations, we can establish the equations related to Chaos Theory and PL, called "*Para*Chaos Equations".

The Certainty Degree (*D*c) obtained by Eq. (5.8) will be on the Chaos Theory/ Paraconsistent: $Dc_{Chaos} = \mu_{Chaos} - \lambda_{Chaos}$

$$Dc_{Chaos} = x_{t-1} - x_{t-1}^2 \tag{5.12}$$

The Contradiction Degree (*D*ct) obtained by Eq. (5.9) on the Chaos Theory/ Paraconsistent will be:

$$Dct_{Chaos} = \mu_{Chaos} + \lambda_{Chaos} - 1 \tag{5.13}$$

$$Dct_{Chaos} = x_{t-1} + x_{t-1}^2 - 1 \tag{5.14}$$

The Chaos Theory/Paraconsistent Logical state ($\varepsilon_{\tau chaos}$) will be represented at the lattice of PAL2v by:

$$\varepsilon_{\tau Chaos(\mu_{Chaos},\lambda_{Chaos})} = (Dc_{Chaos}, Dct_{Chaos}) \tag{5.15}$$

The K-Factor of the Logistic Map in the analysis of Chaos Theory/Paraconsistent will be the multiplicative value that will cause disequilibrium in each Chaos Theory/ Paraconsistent Logical state ($\varepsilon_{\tau chaos}$) generated at each iteration. Therefore, the Certainty Degree times the K disequilibrium factor will give its future value, or the next value in the form of Level of favorable Evidence:

$$\mu_{Chaos_{t+1}} = K.Dc_{Chaos} \tag{5.16}$$

where: $\mu_{Chaos_{t+1}}$ = favorable Evidence Degree resulted in each iteration.

As the Chaos Paraconsistent Logical state ($\varepsilon_{\tau chaos}$) is also formed by the Contradiction Degree, so likewise K disequilibrium Factor will give its future value in the form of unfavorable Evidence Degree:

$$\lambda_{Chaos_{t+1}} = K.Dct_{Chaos} \tag{5.17}$$

The Chaos Theory/Paraconsistent Logical state ($\varepsilon_{\tau chaos}$) resulted in each iteration, can be written as:

$$\varepsilon_{\tau(\mu_{Chaos},\lambda_{Chaos})t+1} = (K.Dc_{Chaos}, K.Dct_{Chaos}) \tag{5.18}$$

In the procedure of the Logistic Map and its action in lattice of PAL2v consists of, from an initial value of an x, consider it as the favorable Evidence Degree and its squared value, the unfavorable Evidence Degree. From these two values it is calculated the Degree of Certainty and the Degree of Contradiction that compose the Chaos Theory/Paraconsistent Logical state ($\Sigma_{\tau chaos}$). Therefore, the values that integrate the Chaos Theory/Paraconsistent Logical state ($\varepsilon_{\tau chaos}$) resulting at each iteration are obtained by:

$$\varepsilon_{\tau(\mu_{Chaos},\lambda_{Chaos})t+1} = (K(\mu_{Chaos} - \lambda_{Chaos}), K(\mu_{Chaos} + \lambda_{Chaos} - 1)) \tag{5.19}$$

After choosing the K disequilibrium Factor, its multiplication by the values of the Evidence degrees and Contradiction Degree provides the future value of favorable Evidence Degree and unfavorable Evidence Degree, which allows to calculate the new Chaos Theory/Paraconsistent Logical state ($\varepsilon_{\tau chaos}$).

Due to the characteristics of the Logistic Map, at each iteration the Chaos Theory/ Paraconsistent Logical state ($\varepsilon_{\tau chaos}$) propagates through a representative Lattice of PAL2v on a trajectory with defined regions, which can be analyzed for the interpretations about the behavior and the balance system.

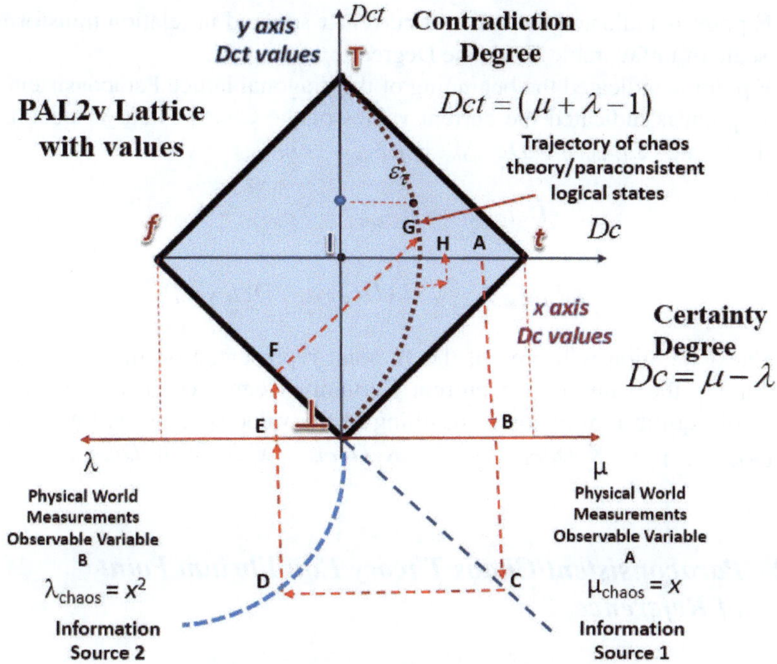

Fig. 5.5 Representation of Evidence Degrees and the trajectory of Chaos Theory/Paraconsistent Logical states $\varepsilon_{\tau Chaos}$ from successive iterations with disequilibrium Factor $K = 1$

The trajectory of the propagating Chaos Theory/Paraconsistent Logical states ($\varepsilon_{\tau chaos}$), due iterations, and how the Evidence degrees are represented in the physical world and their correlations with the Paraconsistent universe are shown in Fig. 5.5.

The sequences of iterations that are made in the analysis of PAL on the Chaos Theory can be studied from the trajectory that starts on line dashed arrows in Fig. 5.5.

The trajectory shows the connection of the representative Lattice of PAL2v with the results obtained by the Logistic Map whose values are represented by Evidence Degrees in the physical world.

The sequence shown in Fig. 5.5 is described as follows:

- At **A** point is represented the value of the Certainty Degree (D_C) before multiplied by the value of K disequilibrium Factor, whose action resulted in the current value of x: $x_t = K.Dc_{Chaos_{t-1}}$.
- At **B** point is indicated the current value x transformed in the scale of favorable Evidence Degree: $x_t = \mu_{Chaos}$.
- At **C** point is indicated the Degree of Evidence value, which is represented by a linear function.
- At **D** point is indicated the current value x squared in relation to the number of iterations in its non-linear characteristic, therefore in the corresponding curve.

- At **E** point is indicated the value of current x squared in relation transformed in the scale of unfavorable Evidence Degree: $x_t^2 = \lambda_{Chaos}$.
- At **F** point is indicated the beginning of the diagonal lattice Paraconsistent.
- At **G** point is indicated the current values of the Chaos Theory/Paraconsistent Logical state $(\Sigma_{\tau chaos}):\ Dc_{Chaos} = \mu_{Chaos} - \lambda_{Chaos}$

$$Dct_{Chaos} = \mu_{Chaos} + \lambda_{Chaos} - 1$$

$$\varepsilon_{\tau(\mu_{Chaos},\lambda_{Chaos})t} = (Dc_{Chaost}, Dct_{Chaost})$$

The next iteration will present the **H** point represented on the horizontal axis of the lattice the value of the current Certainty Degree (D_C) multiplied by the value of disequilibrium factor K, resulting in the value of x in the future: $x_{t+1} = K.Dc_{Chaos_t}\ x_{t+1} = K.Dc_{Chaos_t}x_{t+1} = K.Dc_{Chaos_t}\ x_{t+1} = K.Dc_{Chaos_t}$

4.2.2 *Paraconsistent/Chaos Theory Equilibrium Point of Reference*

In the analysis of PAL2v, the values of Contradiction Degree exposed on the vertical axis of Lattice of PAL2v may be in the range: $-1 \leq Dct_{Chaos} \leq +1$.

However, the sum of the Certainty Degrees and Contradiction Degrees never exceed unity. It is verified that in the study of Paraconsistent Logic in Chaos Theory for a condition of maximum value of the Certainty Degree with the Value of the Degree of Contradiction null, using the Eq. (5.13), we have: $Dct_{Chaos} = \mu_{Chaos} + \lambda_{Chaos} - 1 = 0$, therefore: $Dct_{Chaos} = x_t + x_t^2 - 1 = 0$, solving: $x_t^2 + x_t - 1 = 0$ resulting in:$x_t = 0.618034$.

The value of the Certainty Degree in this condition will be:
$Dc_{Chaos} = x_t - x_t^2 \rightarrow Dc_{Chaos} = 0.618034 - (0.618034)^2$

$$Dc_{Chaos} = 0.236067974$$

For the calculation of the disequilibrium Factor K used for that state in which there is no contradiction, we have:

- the favorable Evidence Degree $x_2 = \mu_{Chaos} = 0.618$
- the Certainty Degree $Dc_{Chaos} = 0.236067974$

From (16) the disequilibrium Factor K is:

$$K = \frac{\mu_{Chaos_{t+1}}}{Dc_{Chaos}} = \frac{0.618}{0.2360} \rightarrow K = 2.618644.$$

All iterations with other disequilibrium Factor K try to establish in the state of total balance of the system. However, although this state is totally balanced it is a transient state, because any infinitesimal variation of the input information represented by x will exit stability. For this Chaos Theory/Paraconsistent Logical state ($\Sigma_{\tau chaos}$), represented by: $\varepsilon_{\tau Chaos} = (Dc_{Chaos}, Dct_{Chaos})$.

$$\varepsilon_\tau = (Dc_{Chaos}, Dct_{Chaos}) \rightarrow Dc_{Chaos} = 0.2360\ Dct_{Chaos} = 0\ \varepsilon_{\tau Chaos} = (0.236067974, 0.0)$$

It turns out that this State Paraconsistent logic of chaos ($\Sigma_{\tau chaos}$), works as point attractor stability for these conditions.

4.3 Results of an Application of the ParaChaos Equations

Table 5.1 shows the results of application of the *Para*Chaos Equations for values of disequilibrium Factor $K = 2.61$, with initial x as 0.1 in a total de 18 iterations.

Table 5.1 Results of application of the *Para*Chaos Equations for values of disequilibrium Factor $K = 2.61$, with initial x as 0.1 in a total de 18 iterations

μ_{Chaos} x	λ_{Chaos} $(x)^2$	Dc_{Chaos} $x-(x)^2$	$D_{ctChaos}$ $x+(x)^2-1$	$\varepsilon_\tau 1(\mu_{Chaos}, \lambda_{Chaos})$	x_{t+1}
0.1	0.01	0.09	−0.8900000	a	0.2349000
0.2349	0.0551780	0.1797219	−0.7099219	b	0.4690743
0.46907439	0.2200307	0.2490436	−0.3108948	c	0.6500038
0.65000381	0.4225049	0.2274988	+ 0.0725087	d	0.5937720
0.59377201	0.3525652	0.2412006	−0.0536627	e	0.6295497
0.62954977	0.3963329	0.2332168	+ 0.0258826	f	0.6086959
0.60869599	0.3705108	0.2381185	−0.0207931	g	0.6216633
0.62166331	0.3864652	0.2351980	+ 0.0081286	h	0.6138668
0.61386687	0.3768325	0.2370343	−0.0093005	i	0.6186596
0.61865961	0.3827397	0.2359198	+ 0.0013993	j	0.6157509
0.61575092	0.3791492	0.2366017	−0.0050998	l	0.6175304
0.61753049	0.3813439	0.2361865	+ 0.0112559	m	0.6164469
0.61644698	0.3800068	0.2364401	−0.00354613	n	0.6171086
0.61710866	0.3808231	0.2362855	−0.00068238	o	0.6167053
0.61670531	0.3803254	0.2363798	−0.00296924	p	0.6169514
0.61680134	0.3804439	0.2363574	−0.00275474	q	0.6168929
0.61689293	0.3805568	0.2363360	−0.00255018	r	0.6168370
0.61683707	0.3804879	0.2363490	−0.00267495	s	0.6168711

The value of $K = 2.61$ was chosen because the application of the *Para*Chaos Equations shows the Paraconsistent/Chaos Theory equilibrium point of reference in which the Certainty Degree is around 0.2360 and the Contradiction Degree is near zero.

In the application of PL in Chaos Theory it is verified that any other iteration in which the values of the disequilibrium Factor K are different from 2.61 are attempts to achieve this Chaos Theory/Paraconsistent Logical state ($\varepsilon_{\tau Chaos}$) which is characterized by null contradiction.

Figure 5.6 shows the Paraconsistent/Chaos Theory Equilibrium Point of Reference with values presented on Table 5.1.

The values presented in Table 5.1 show a relationship of the *Para*Chaos equations with the golden ratio geometry. For State PL of chaos ($\varepsilon_{\tau chaos}$), that is a point attractor stability represented by: $\varepsilon_{\tau Chaos} = (Dc_{Chaos}, Dct_{Chaos})$

$$Dc_{Chaos} = 0.2360$$

$$Dct_{Chaos} = 0$$

$$\mu_{Chaos_{t+1}} = 0.616871147$$

This value is very close to the known gold mean that is: $\phi = \frac{\sqrt{5}-1}{2}\phi = \frac{\sqrt{5}-1}{2}$.

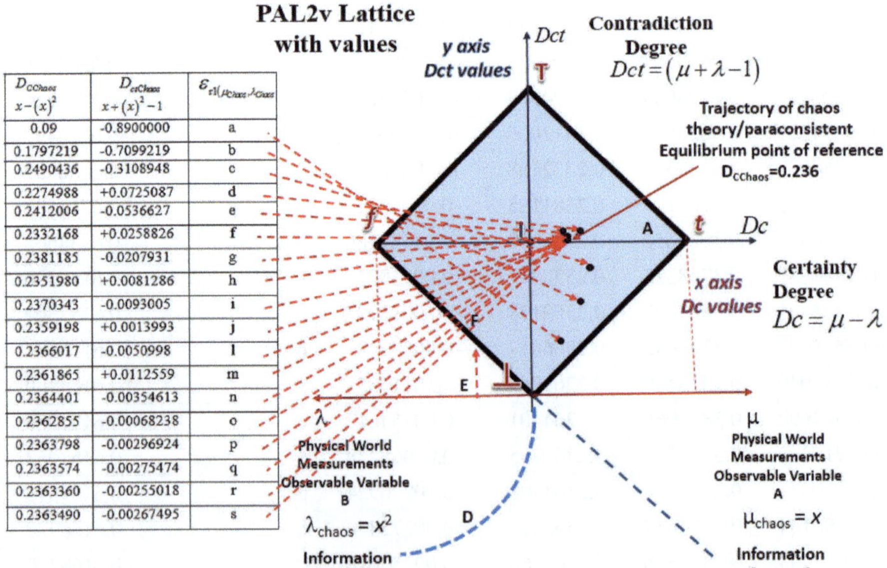

D_{CChaos} $x-(x)^2$	$D_{ctChaos}$ $x+(x)^2-1$	$\varepsilon_{\tau 1}(\mu_{Chaos}, \lambda_{Chaos})$
0.09	-0.8900000	a
0.1797219	-0.7099219	b
0.2490436	-0.3108948	c
0.2274988	+0.0725087	d
0.2412006	-0.0536627	e
0.2332168	+0.0258826	f
0.2381185	-0.0207931	g
0.2351980	+0.0081286	h
0.2370343	-0.0093005	i
0.2359198	+0.0013993	j
0.2366017	-0.0050998	l
0.2361865	+0.0112559	m
0.2364401	-0.00354613	n
0.2362855	-0.00068238	o
0.2363798	-0.00296924	p
0.2363574	-0.00275474	q
0.2363360	-0.00255018	r
0.2363490	-0.00267495	s

Fig. 5.6 Representation of paraconsistent/chaos theory equilibrium point of reference with values presented on Table 5.1

4.3.1 Computer Simulations Results

We present the results of computer simulations made with the *Para*Chaos equations where were varied some values.

In the Fig. 5.7 the disequilibrium Factor *K* was kept constant with 2.61 and with three different initial evidence degrees; $\mu_{1initial} = 0.1$, $\mu_{2initial} = 0.25$ and $\mu_{3initial} = 0.5$. Note that the value of the degree of certainty reaches the Equilibrium Point of Reference (point attractor stability) with less number of iterations using the lowest degree of initial evidence,$\mu_{1initial} = 0.1$.

In the Fig. 5.8 the value of the initial evidence degree of was kept constant $\mu_{3initial} = 0.1$ with three different values of disequilibrium Factor *K;* $K_1 = 2.61$, $K_2 = 2.55$ and $K_3 = 2.75$.

Note that the value of the degree of certainty reaches the Equilibrium Point of Reference (point attractor stability) with less number of iterations using the lowest disequilibrium Factor *K*, $K_1 = 2.61$.

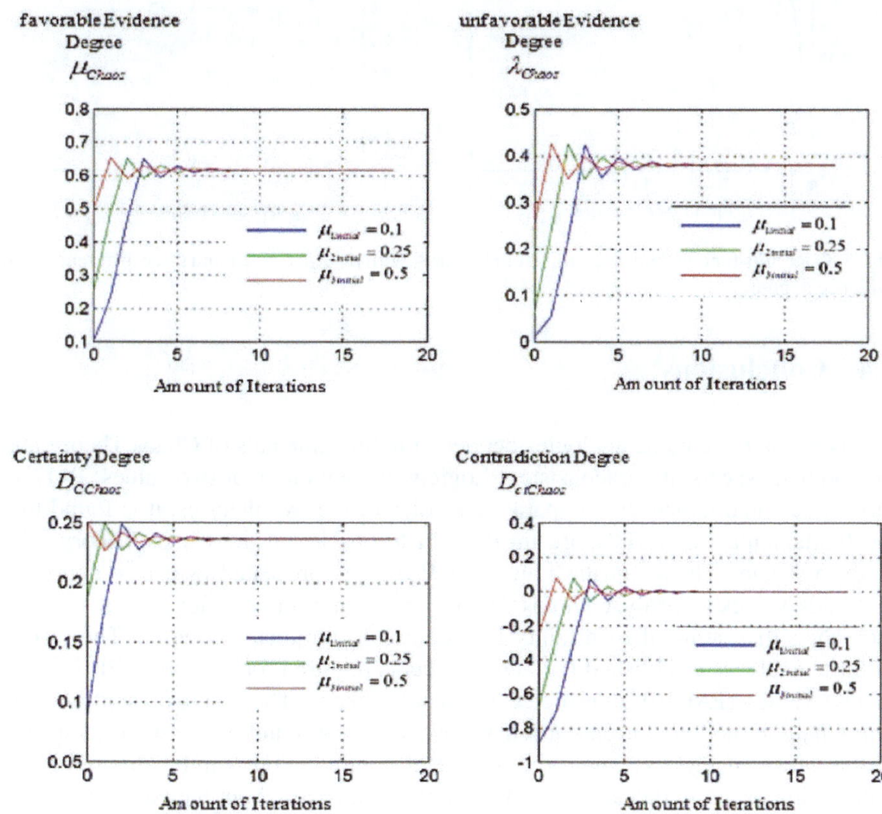

Fig. 5.7 Results for constant initial evidence degree with 2.61 and three different initial evidence degrees

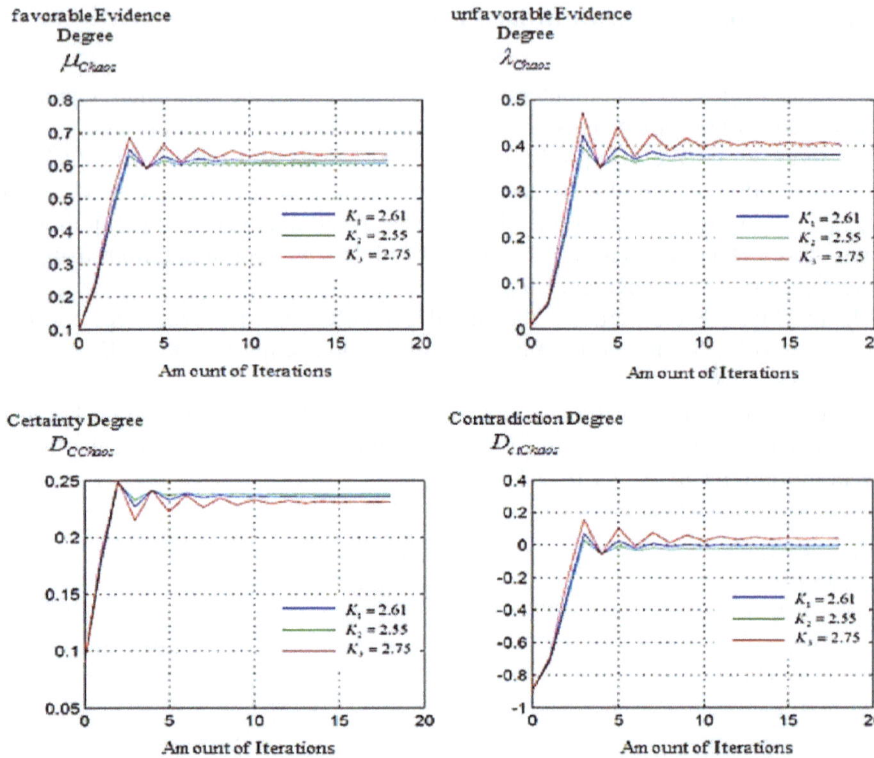

Fig. 5.8 Results for constant disequilibrium Factor K with $\mu_{1initial} = 0.1$ and three different initial evidence degrees

4.4 Conclusions

In this chapter, we made analogies between the fundamentals of Chaos Theory and interpretive aspects of Paraconsistent Logic with annotation of two values (PAL2v) in its associated lattice. In comparative studies of the two theories, it is found that the fundamentals of PAL2v are intrinsic in the basics of the Chaos Theory. The study of Chaos Theory in the Lattice of PAL2v is presented as a new approach that adds values to analyze Chaos Theory in the area of complex systems due to the values that define the Certainty Degree and Contradiction Degree. The results of these analyzes established the main concepts required for the use of PL in form of two values (PAL2v), interpreted in Chaos Theory. The result of applying the *Para*Chaos Equations into a particular case shows an equilibrium point identified in the lattice of PAL2v, where the degree of contradiction is null. This reference point is important in deductions and behavioral analyzes about the balance state of dynamical systems that will further be studied. We also highlight the fact that the value of the degree of evidence favorable to be strongly related to the Golden Ratio of geometry. In that work, comparisons of these values will identify the system behavior in the passages by bifurcation points and the boundaries between phases defining the stability and the states of the chaotic system. It is verified that the PL states that present themselves in every interaction allow better visualization of the results, which facilitate comparisons over the balance of the system analyzed from the variations in K factor. The basics and the equations derived from this study open interesting and new ways for direct applications of the two theories in different fields of knowledge, mainly dealing with control and stability analysis of dynamic systems. Furthermore, the Dct_{Chaos} can be seen as an important indicative factor of disequilibrium in the chaos analysis using the PAL2v.

References

1. Bar-Yam, Y.: Dynamics of complex systems. Addison-Wesley, 1st edn (1997)
2. Kolmogorov, A.N.: Three approaches to the quantitative definition of information. Probl. Inform. Transm. **1**, 1–7 (1965)
3. Adamatzky, A.: Identification of cellular automata. Taylor & Francis Ltd., 1st edn (1994)
4. Langton, C.G.: Computation at edge of chaos: phase transitions and emergent computation. Physica D **42**, 12–37 (1990)
5. Kolmogorov, A.N.: Logical basis for information theory and probability theory. IEEE Trans. Inform. Theory **14**, 662–664 (1968)
6. Ott, E.: Chaos in dynamical systems, 2nd edn. Cambridge University Press (2002)
7. Stradner, J., et al.: Algorithmic requirements for swarm intelligence in differently coupled collective systems. Elsevier/ Chaos, Solit. Fractals. **50**, 100–114 (2013)
8. Boccaletti, S., et al.: The control of chaos: theory and applications. Elsevier Sci. Phys. Rep. **329**, 103–197 (2000)
9. Alligood, K.T., Sauer, T.D., Yorke, J.A.: Chaos an introduction to dynamical systems. Springer-Verlag, New York, Inc (1997)
10. Devaney, R.: An introduction to chaotic dynamical systems, 2nd edn. Addison-Wesley, Redwood City (1989)
11. Da Silva Fiho, J.I., Lambert-Torres, G., Abe, J.M.: Uncertainty treatment using paraconsistent logic: introducing paraconsistent artificial neural networks, p. 328. IOS Press, Amsterdam (2010)
12. Da Silva Filho, J.I.: Paraconsistent annotated logic in analysis of physical systems: introducing the para quantum factor of quantization h_{ψ}. J. Mod. Phys. **2**, 1397–1409 (328). Published Online. https://doi.org/10.4236/jmp.2011.211172
13. Abe, J.M., Da Silva Filho, J.I.: Inconsistency and electronic circuits. In: Alpaydin, E. (eds.) Proceedings of EIS'98 International ICSC Symposium on Engineering of Intelligent Systems. Artificial Intelligence, ICSC Academic Press, Rochester, vol. 3, pp. 191–197 (1998)
14. Lorenz, E.N.: The essence of Chaos, 1.0 Edition. University of Whashington Press, p. 240
15. Malthus, T.R.: An essay on the principle of population, as it affects the future improvement of society, with remarks on the speculations of Mr. Godwin, M. Condorcet, and other Writers, 1st edn. London, Johnson (1798)
16. Verhulst, P.F.: Notice sur la loi que la population poursuit dans son accroissement. Corresp. Math. Phys. **10**, 113–121 (1838)
17. May, R.M.: Simple mathematical models with very complicated dynamics. Nature **261**, 459–467 (1976)
18. Lorenz, E.N.: Deterministic non periodic flow. J. Atmos. Sci. **20**, 130–141 (1963)

19. Feigenbaum, M.: Quantitative universality for a class of nonlinear transformations. J. Stat. Phys. **19**, 25–52 (1978)
20. Hahn, W.: Theory and application of Liapunov's direct method. Prentice-Hall, Englewood Cliffs, NJ (1963)
21. Da Silva Filho, J.I., Rocco, A.: Power systems outage possibilities analysis by Paraconsistent Logic. Power and Energy Society General Meeting - Conversion and Delivery of Electrical Energy in the 21st Century, pp. 1–6, IEEE ISBN: 978–1–4244–1905–0, ISSN: 1932–5517. Pittsburgh, PA (2008)

Chapter 5
A Paraconsistent Artificial Neural Cell of Learning by Contradiction Extraction ($PANC_{LCTX}$) with Application Examples

Arnaldo de Carvalho Jr., João Inácio Da Silva Filho, Márcio de Freitas Minicz, Gustavo R. Matuck, Hyghor Miranda Côrtes, Dorotéa Vilanova Garcia, Paulo Marcelo Tasinaffo, and Jair Minoro Abe

Contents

Abstract Paraconsistent logic (PL) allows decision-making systems to deal with uncertainty and contradictory signals offering an alternative to the classical logic.

A. de Carvalho Jr.
Instituto Federal de Educação, Ciência e Tecnologia de São Paulo –IFSP, Maria Cristina Street, 50, Cubatão City, São Paulo 11533-160, Brazil
e-mail: adecarvalhojr@ifsp.edu.br

A. de Carvalho Jr. · J. I. Da Silva Filho (✉) · H. M. Côrtes · D. V. Garcia
Laboratory of Applied Paraconsistent Logic, Santa Cecília University, Unisanta, Oswaldo Cruz Street, 288, Santos City, São Paulo 11045-000, Brazil
e-mail: inacio@unisanta.br

M. de Freitas Minicz · G. R. Matuck · P. M. Tasinaffo
Instituto Tecnológico de Aeronautica (ITA), Division of Computer Science, Marechal Eduardo Gomes Square, 50, São José Dos Campos City, São Paulo 12.228-900, Brazil
e-mail: matuck@ita.br

P. M. Tasinaffo
e-mail: tasinaffo@ita.br

J. M. Abe
Graduate Program in Production Engineering, Paulista University, São Paulo, Brazil

J. M. Abe (ed.), *Advances in Applied Logics*, Intelligent Systems Reference Library 243,
https://doi.org/10.1007/978-3-031-35759-6_5

Using the interpretations and equations of the PL, it is possible to program blocks of codes called Paraconsistent Artificial Neural Cells (PANCs) used to design data processing systems through paraconsistent artificial neural networks (PANnets). Among the several types of paraconsistent cells, one of the most used is the Paraconsistent Artificial Neural Cell of learning ($PANC_L$). In this chapter, we present a new form of the $PANC_L$ based on extraction of the contradiction effects between the input and the previous output, called $PANC_{LCTX}$. The results presented here demonstrate that the proposed cell works effectively integrating asymptotic mode values, with practical applications for industry in signal analysis, estimation and treatment.

Keywords Paraconsistent logic · Paraconsistent annotated logic · Paraconsistent artificial neural cell · Algorithm

5.1 Introduction

In some real-world situations, classical logic, which is supported by strictly binary principles, fails to provide foundations for building algorithms capable of responding properly. In situations where information is uncertain or contradictory, classical logic becomes inoperative. In order to obtain systems capable of alleviating this problem, it is necessary to use other types of logic whose foundations are opposed to or challenge the binary laws of classical logic.

The logics that were created for this purpose are called non-classical logics. Paraconsistent Logic is a logic belonging to the family of non-classical logics and has in its foundation the property of supporting signals with contradictory information without overriding the conclusions [1, 2]. Through linear transformations and interpretations of values established in a four vertices Lattice (Hasse diagram) associated to a PL, called Paraconsistent Annotated Logic (PAL), it is possible to elaborate learning equations in paraconsistent analysis nodes (PAN), with results that simulate the behavior of a biological neuron [3, 4]. With studies based on the PAN, it was possible to build Paraconsistent Artificial Neural Cells (PANC) [5]. When interconnected, PANC form Paraconsistent Artificial Neural networks (PANnet), taking full advantage of the analytical capacity of PL [5–7].

Differently from artificial neural networks (ANNs), typically based on just one type of cell, there are several types of cells that can be used in PANnet [3]. Among the several existing types of cells, the Paraconsistent Artificial Neural Cell of Learning ($PANC_L$) stands out. This cell has the ability to learn input pattern with the number of iterations defined by a Learning Factor (F_L), that varies in the closed range between [0,1] and belonging to the set of Real numbers [8–10]. The properties and responses offered by $PANC_L$ allows its application in the network or at work standalone with very useful applications in the industry automation [8–10]. However, due to its structure, there are restrictions in your answers for some configurations.

Based on the constraints of the $PANC_L$, in this study we propose a different algorithm for the learning cell. We will call this new configuration PANCL by contradiction effect extraction ($PANC_{LCTX}$). This new cell is capable of learning and unlearning pattern value presented at its input at controlled speeds.

In addition to this introduction, the text is organized as follows: Sect. 5.2 presents an introduction of the Paraconsistent Logic (PL), the Paraconsistent Annotated Logic with annotation of two values (PAL2v) and its main properties. In addition, an overview of the Standard Paraconsistent Artificial Neural cell (sPANC) and a deep study of the PANCL including some issues are presented in this section. The $PANC_{LCTX}$, its equations, and how it solves the issues of the $PANC_L$ are presented in Sect. 5.3. Results of application examples of the $PANC_{LCTX}$ are presented at the Sect. 5.4. Section 5.5 concludes with the considerations about the $PANC_{LCTX}$ based on the results of Sect. 5.4.

5.2 Paraconsistent Logic (PL)

The classical logic is based on stringent and inflexible binary laws not admitting situations of redundancy, inconsistencies or incomplete data [1, 3, 4]. There is no contradiction in classical logic. This means that when dealing with the same context, something cannot be both true and not true at the same time. In order to act in situations where the application of binary logic is not proper, other types of logic, called non-classical, have been created [1–3]. Within the many ideas for non-classical logics, there has emerged a family of logics that presents as its main theoretical foundation the revocation of the non-contradiction principle, called Paraconsistent Logics [7, 10].

When a lattice with logical states represented at its vertices is associated with PL, it is called Paraconsistent Annotated Logic (PAL). In this PAL-representation, propositions can be formulated, and analyzed through mathematical interpretations in a lattice with four vertices (Hasse Lattice), for example. The four extreme logic states represented at the vertices of the PAL lattice are True (t), False (F), Inconsistent (T) and Paracomplete ($\perp$).

Paraconsistent Annotated Logic with annotation of two values (PAL2v) is part of the PL family of logics, where a proposition P is analyzed by evidence from 2 variables, represented in a Hasse Lattice [1, 5, 6]. An operator " ~ " is introduced, defined as [2, 6]: $\sim = |\tau| \rightarrow |\tau|$ where $\tau = \{(\mu, \lambda)\} | \mu, \lambda | \in [0, 1] \subset \mathbb{R}$.

The evidences that support the proposition P are annotated in a pair of ordered values (μ_1, μ_2), labelled the degree of favorable evidence (μ or μ_1) and degree of unfavorable evidence (λ) complementary to the second input (μ_2) as (1) [5, 7].

$$\lambda = (1 - \mu_2) \tag{5.1}$$

Figure 5.1 presents the lattice diagram for PAL and PAL2v [5].

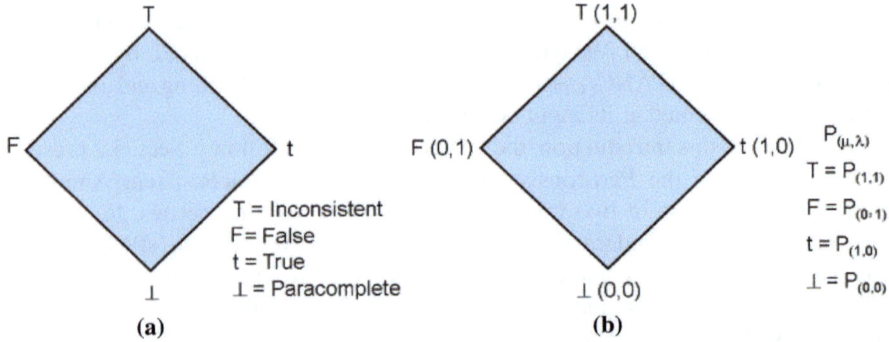

Fig. 5.1 Associated Lattice of Paraconsistent Annotated Logic. **a** Hasse finite lattice with paraconsistent extreme states. **b** PAL2v-Lattice with annotations

Based on the μ and λ inputs, it is possible to calculate the values of the certainty degree (*Dc*) and contradiction degree (*Dct*) to the proposition, according to Eqs. (5.2) and (5.3).

$$Dc = \mu - \lambda \tag{5.2}$$

$$Dct = \mu + \lambda - 1 \tag{5.3}$$

The point ετ(*Dc*, *Dct*) is the paraconsistent logical state, in the lattice diagram, as Fig. 5.2, function of the inputs μ and λ [5].

For a proposition *P*, the limits of the pair (μ, λ) indicates:

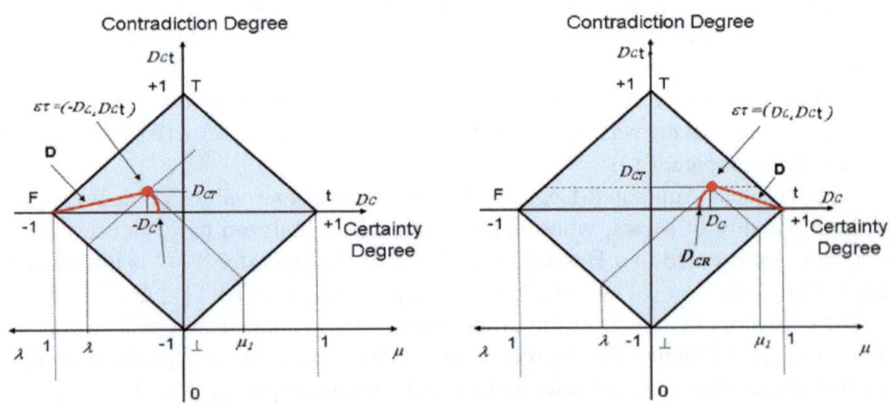

Fig. 5.2 Lattice of PAL2v, with input evidences (μ,λ) and projection of paraconsistent logical state, ετ(D_C,*Dct*)

(1.0) → maximum favorable evidence and unfavorable evidence null, meaning an absolute true (*t* corner, H axis) logical connotation for *P*. The point ετ(*Dc*, *Dct*) is also equal to (1,0).
(0,1) → favorable evidence null and maximum unfavorable evidence, giving a connotation of logical falsity (F corner, H axis) for *P*. The point $\varepsilon\tau(D_C, Dct)$ is equal to (-1,0).
(1,1) → maximum favorable and unfavorable evidences, attributing a logical connotation of inconsistency (T corner, V axis) to *P*. The point $\varepsilon\tau(D_C, Dct)$ is equal to (0,1).
(0,0) → favorable and unfavorable evidence null, assigning a logical connotation of paracomplete (⊥ corner, V axis), for ***P***. The point $\varepsilon\tau(D_C, Dct)$ is also equal to (0,-1).

The results can be more accurate [6], by extracting the contradiction effects by successive analysis, calculating the degree of real certainty (Dc_R) by projecting the line segment D (4), to the horizontal X (5), as presented by Fig. 5.2.

$$D = \sqrt{(1 - |Dc|)^2 + Dct^2} \tag{5.4}$$

$$If\ Dc > 0 \rightarrow\ Dc_R = 1 - D;\ \ If\ \ Dc < 0 \rightarrow\ Dc_R = D - 1 \tag{5.5}$$

5.2.1 *Paraconsistent Artificial Neural Cell*

PAL2v interpretation and formulas can be implemented as an algorithm for processing devices [2]. One of the widely used PAL2v algorithms is called paraconsistent analysis node (PAN), whose symbol is presented in Fig. 5.3. PAN studies have resulted in the creation of a Paraconsistent Artificial Neural Cell (PANC) [7] that is a block of code that performs specific paraconsistent analysis. The standard PANC is a code block that has all the paraconsistent rules implemented and its symbol is presented in Fig. 5.4 [5, 7].

Dc, *D*ct and Dc_R outputs have values between {-1,1}. In order to allow the interconnection of PAN or PANCs, making Paraconsistent Artificial Neural Networks (PAN_{NET}), these values must be normalized back to limits {0,1} as Eqs. (5.6) to (5.8) and called respectively resulting evidence degree (μ_E), resulting real evidence degree and resulting contradiction degree (μ_{ct}) outputs [5, 7].

$$\mu_E = \frac{Dc + 1}{2} = \frac{\mu - \lambda + 1}{2} \tag{5.6}$$

$$\mu_{ER} = \frac{Dc_R + 1}{2} \tag{5.7}$$

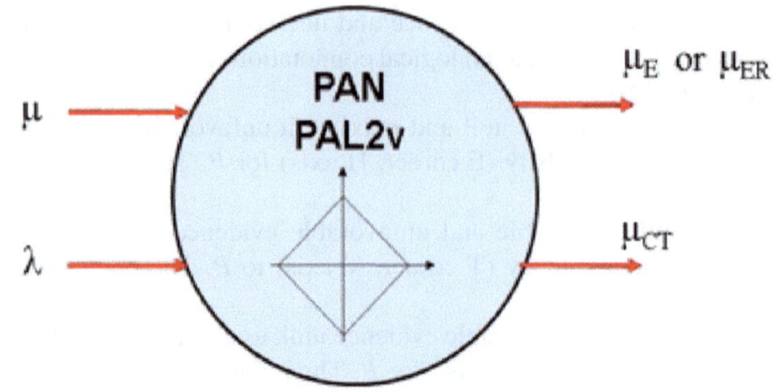

Fig. 5.3 Symbol of Paraconsistent Analysis Node (PAN)

Fig. 5.4 Representation of Standard Paraconsistent Artificial Neural Cell (sPANC)

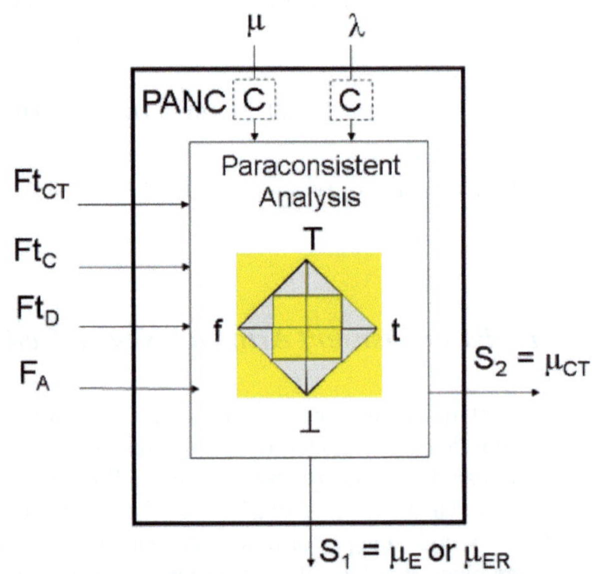

$$\mu_{ct} = \frac{Dct + 1}{2} = \frac{\mu + \lambda}{2} \tag{5.8}$$

The contradiction tolerance factor (Ft_{CT}), certainty tolerance factor (Ft_C), decision tolerance factor (Ft_D), and learning factor (F_L) are optional inputs in order to allow different limit values between {0,1} to *Dc*, *Dct*, decision and learning rate [5].

5.2.2 Paraconsistent Artificial Neural Cell of Learning

Paraconsistent Artificial Neural Cell of Learning—$PANC_L$ can be designed by applying the complement of the previous output to the current unfavorable degree of evidence (λ) input, and executing a series of iterations. The cell output tends to follow, or "learn", the signal of the favorable degree of evidence (μ), reason why it is called Paraconsistent Artificial Neural Cell of learning, whose symbol is presented in Fig. 5.5. The output of resulting evidence degree (μ_E) is then calculated as (9) or (10) [5].

Equation (5.9) applies the learning factor F_L only to the unfavorable evidence (λ) while Eq. (5.10) applies F_L for both favorable (μ) and unfavorable (λ) evidences.

$$\mu_{Ek} = \frac{\{\mu_1 - \mu_{E(K-1)C}.F_L\} + 1}{2} = \frac{\{\mu_1 - (1 - \mu_{E(K-1)}).F_L\} + 1}{2} \quad (5.9)$$

$$\mu_{Ek} = \frac{\{\mu_1 - \mu_{E(K-1)C}.F_L\} + 1}{2} = \frac{\{\mu_1.F_L - (1 - \mu_{E(K-1)}).F_L\} + 1}{2} \quad (5.10)$$

These two equations respond well to the behavior of a biological neuron whenever the learning factor (F_L) is equal to 1, regardless of the value of the input pattern μ.

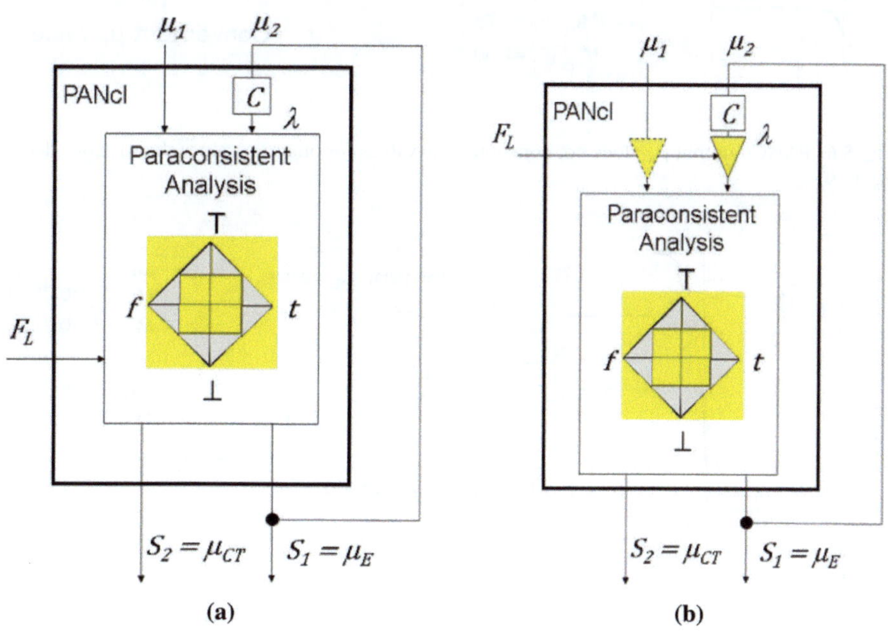

Fig. 5.5 Cells representations. **a** Representation of Standard Paraconsistent Artificial Neural Cell (sPANC). **b** Paraconsistent Artificial Neural Cell of Learning $PANC_L$ with Factor F_L applied to the inputs

However, when it is necessary to vary both the F_L and the input μ, Eq. (5.9) converges to a value different from the input, as presented by Fig. 5.6.

Evaluating Eq. (5.10), the output also does not converge to input μ for other F_L values than 1, as presented by Fig. 5.7. Also, when F_L is equal to zero, regardless of the input μ, the resulting evidence degree (μ_E) at the cell output is always 0.5.

Considering Eq. (5.10), it was expected that in this situation the output would be a step function, jumping directly to the pattern to be learned. In order to adapt the output signals to these special situations in which Eqs. (5.9) and (5.10) do not respond to the expected behavior for some applications, an improvement of the $PANC_L$ is necessary [5, 7].

The improved cell to solve these issues is presented at the next section.

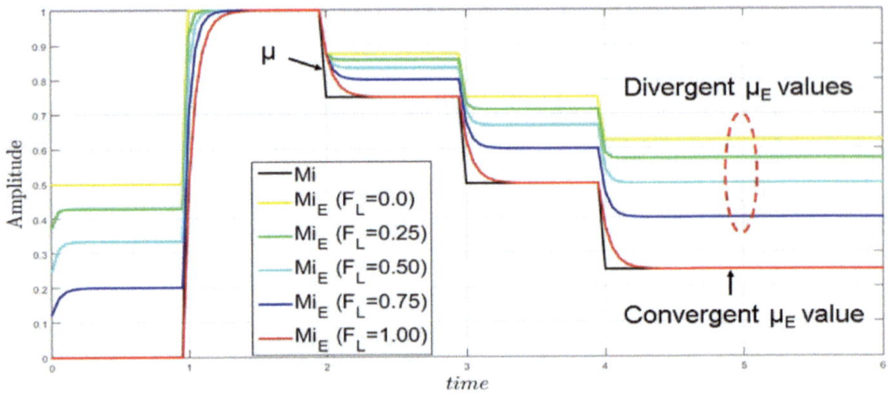

Fig. 5.6 $PANC_L$ output μ_E does not converge to μ values for other F_L values than 1, considering Eq. (5.9)

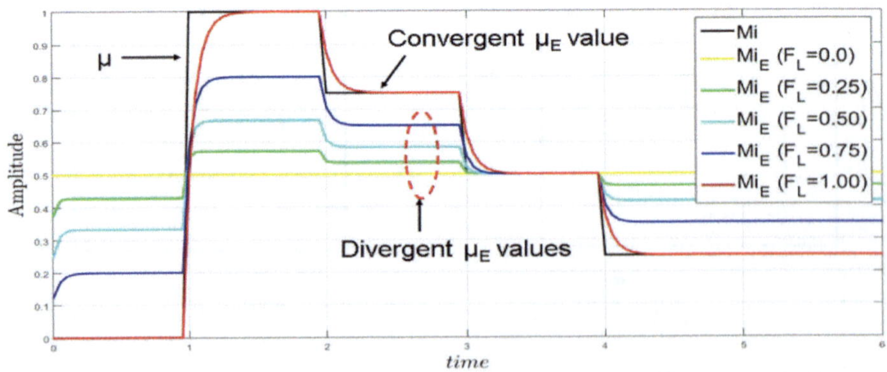

Fig. 5.7 $PANC_L$ output μ_E does not converge to μ values for other F_L values than 1, considering Eq. (5.10)

5.3 Paraconsistent Artificial Neural Cell of Learning by Contradiction Extraction

The cell we propose in this article starts with the contradiction degree (*D*ct) integrating the iterations *k*, as Eq. (5.11), and uses the Eq. (5.12) to calculate the current resulting evidence degree (μ_E). As the proposed cell reduces the degree of contradiction to each iteration, it was named the Paraconsistent Artificial Neural Cell of learning by contradiction effect extraction ($PANC_{LCTX}$).

$$\mu_{EK} = \mu_{E(K-1)} + Dct.F_L \tag{5.11}$$

$$Dct = \mu + \mu_{E(K-1)C} - 1 \rightarrow Dct = \mu + 1 - \mu_{E(K-1)} - 1$$

$$Dct = \mu - \mu_{E(K-1)} \tag{5.12}$$

The learning factor must be in the range $0 < F_L < 1$, where:

(a) $F_L = 0.0 = >$ there is no cell learning;
(b) $F_L = 0.5 = >$ same learning as standard $PANC_L$ with unitary F_L;
(c) $F_L = 1.0 = >$ learning immediately, μ_E is equal to favorable evidence (μ).

The $PANC_{LCTX}$ can be explained as an adaptation on the standard PANC, taking the Eq. (5.11) and breaking down to the Eqs. (5.13) to (5.15), for a sample number k. Figure 5.8 presents the construction of $PANC_{LCTX}$ from the standard PANC.

$$\mu_{EK} = \mu_{E(k-1)} + Dct.F_L = \mu_{E(k-1)} + \left(\mu_k - \mu_{E(k-1)}\right).F_L \tag{5.13}$$

$$\mu_{EK} = \mu_{E(k-1)} + \mu_k.F_L - \mu_{E(k-1)}.F_L \tag{5.14}$$

$$\mu_{EK} = \mu_k.F_L - \mu_{E(k-1)}.(1 - F_L) \tag{5.15}$$

Unlike the results of the Figs. 5.6 and 5.7, the output μ_E is capable of learning the value applied to the input μ even using different F_L values.

With lower F_L values, more samples are required in order for the output μ_E to converge to the value of μ. Also, for F_L equal to 1, the output μ_E is exactly the same as the input μ, without any delay.

The initial value of μ_E is 0.5, and based on Eq. (5.17), for F_L equal to zero, there is no learning. In this case $\mu_E[n]$ is equal to $\mu_E[n\text{-}1]$, until $\mu_E[0]$, so the output is fixed in 0.5. This is recognized by the foundations of the PAL2v as a logical state undefined I (corresponding to $D_C = 0$).

Figure 5.9 presents the output μ_E of the $PANC_{LCTX}$ as the input μ changes, for different values of F_L.

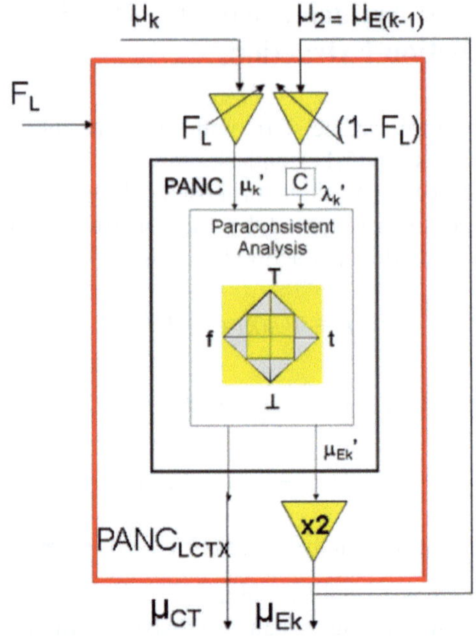

Fig. 5.8 $PANC_{LCTX}$ construction from the standard PANC

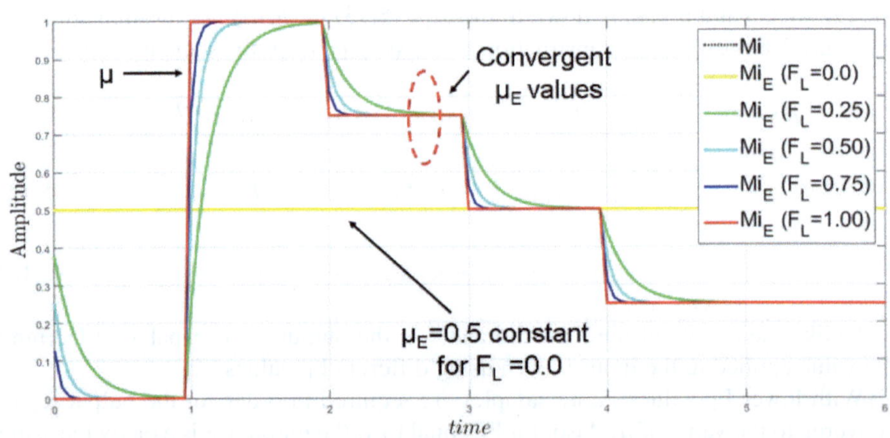

Fig. 5.9 $PANC_{LCTX}$ output μ_E converges to μ values, even considering different F_L values

The Table 5.1 presents the number of samples required by the output μ_E of the $PANC_{LCTX}$ to learn the input μ, for different values of F_L.

With respect to the speed or the number of iterations, the convergence is influenced by the Learning Factor F_L.

Table 5.1 Iterations required by $PANC_{LCTX}$ to learn μ relative to *FL*

FL	Iterations
1.00	1
0.90	4
0.75	6
0.50	11
0.25	26
0.10	68
0.00	Undefined

The output μ_{ct} always converges to value 0.5 regardless the F_L value, except to F_L equal to zero, when the output μ_{ct} is $(\mu + 0.5)/2$. As closer to 1.0 is the learning factor, closer to an ideal impulse is the output μ_{ct}.

Figure 5.10 presents the output μ_{ct} as the input μ changes, for different values of F_L.

In some neural network applications, it may be interesting that the cell uses a number of iterations to learn the pattern of truth and another to unlearn, that is, to learn the pattern of falsehood. This can be easily accomplished using a learning factor (F_L) and an unlearning factor (F_{UL}) with the algorithm presented below. This allows better behavior control of the cell.

Step 1 $\mu_E[0] = 0.5$;
Step 2 Define F_L value;
Step 3 Define F_{UL} value;
Step 4 Enter the value of $\mu[n]$;
Step 5 $\lambda[n] = 1 - \mu_E[n-1]$;
Step 6 $Dct = \mu[n] - \lambda[n] + 1$;
Step 7 If $\mu[n] > \mu_E[n-1]$, then $\mu_E[n] + Dct^*F_L$;
Step 8 If $\mu[n] < \mu_E[n-1]$, then $\mu_E[n] + Dct^*F_{UL}$;

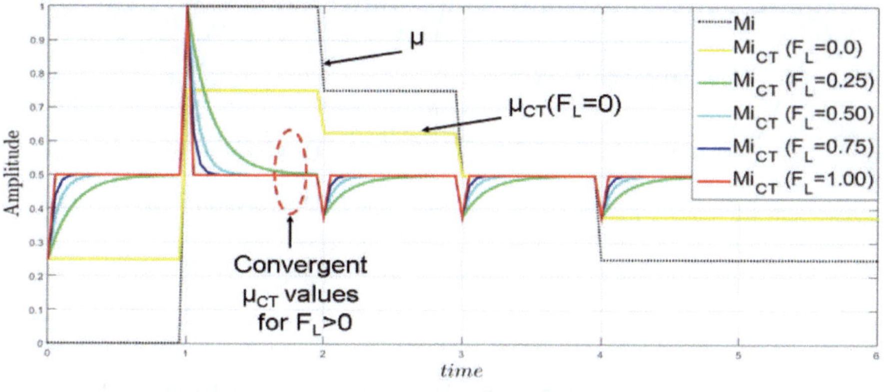

Fig. 5.10 $PANC_{LCTX}$ output μ_{ct} in function of μ changes, for different F_L valucs

Step 9 Go to Step 4;

The F_L acts when the input pattern μ is greater than the evidence degree μ_E and the F_{UL} in the inverse situation, that is, when the evidence degree μ_E is greater than the input pattern μ.

Figure 5.11 shows the learning and unlearning process of the $PANC_{LCTX}$.

The same number of iterations the $PANC_{LCTX}$ takes to learn one value presented at the input μ is the required to learn a new one. If a value of 1 is presented at the input μ, representing the pattern of truth and, after the process of learning of the $PANC_{LCTX}$ be completed a new value of 0 is presented at the input μ, representing the pattern of falsehood, the cell requires the same number of iterations to learn the pattern of falsehood, or unlearn the pattern of truth.

For practical uses, it is noted that he output μ_E of the $PANC_{LCTX}$ can be compared to a discrete first order low pass filter [17], where there is a relation between the learning factor, sampling rate (T_S) and time constant (τ) of the filter (16).

$$F_L = \frac{T_S}{T_S + \tau}; \quad (1 - F_L) = \frac{\tau}{T_S + \tau} \tag{5.16}$$

Making the equations in function of sample *n*, from the Eq. (5.15) the output μ_E is as presented at Eq. (5.17). It is possible to calculate the transfer function of the $PANC_{LCTX}$ for output μ_E and to apply the *Z* transform, whose result is presented at Eq. (5.18). Also, from Eq. (5.8), for the $PANC_{LCTX}$ the output μ_{ct} is as presented at Eq. (5.19) and the *Z* transform of the transfer function is presented at Eq. (5.20).

$$\mu_E[n] = F_L.\mu[n] + (1 - F_L).\mu_E[n-1] \tag{5.17}$$

$$\frac{\mu_E(z)}{\mu(z)} = F_L.\left(\frac{z}{z - (1 - F_L)}\right) \tag{5.18}$$

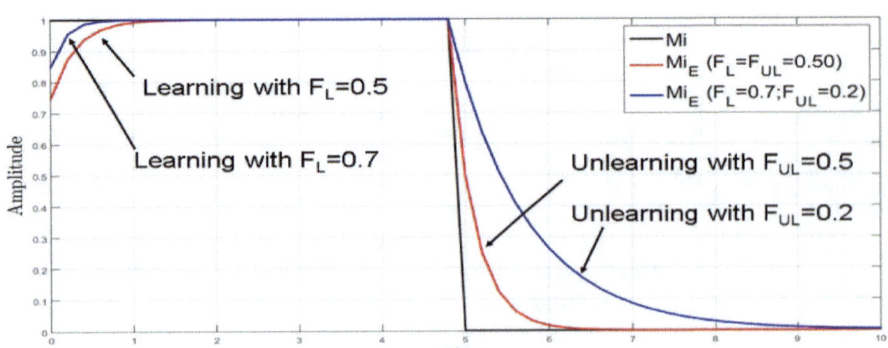

Fig. 5.11 $PANC_{LCTX}$ learning and unlearning the pattern of truth, for different F_L values

$$\mu_{CT}[n] = \frac{1}{2}(\mu[n] - \mu[n-1) + (1 - F_L) * \left(\mu_{CT}[n-1] - \frac{1}{2}\right) + \frac{1}{2} \quad (5.19)$$

$$\frac{\mu_{CT}(z)}{\mu(z)} = \frac{1}{2} * \left(1 + \frac{F_L z^{-1}}{1 - (1 - F_L)z^{-1}}\right) \quad (5.20)$$

Based on the Eq. (5.18) the $PANC_{LCTX}$ works as a type of integrator or 1st order low pass filter (LPF) when using the output μ_E and as a type of differentiator or, based on Eq. (5.20), as 1st order high pass filter (HPF) when using the output μ_{ct}.

5.4 Application Examples in the Industry

Since the transfer function of the $PANC_{LCTX}$ works as a type of integrator or differentiator, as Eqs. (5.18)-(5.20) presented earlier at the session 3 of this study, it is possible to use the cell alone or in a network with the purpose of treatment and filtering of variables used in the industry.

5.4.1 Variable Estimator Configured with $PANC_{LCTX}$

The Kalman filter design example available in (THE MATHWORKS INC., 2019) [11, 12] is used here for a comparison with $PANC_{LCTX}$ as estimator. The example presents a plant of and Kalman, according to Fig. 5.12, to which a $PANC_{LCTX}$ (in red) now is added (in red).

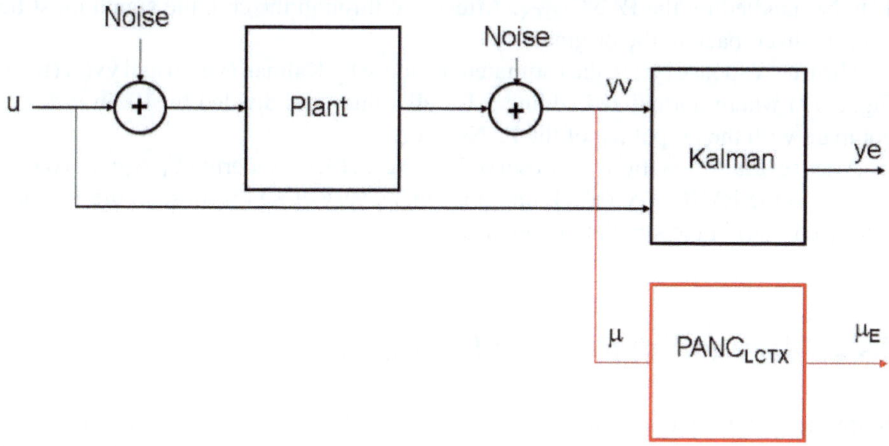

Fig. 5.12 Example of Kalman Estimator and $PANC_{LCTX}$ in Automation Plant. *Source* Mathworks, Kalman Filter Design (2019)

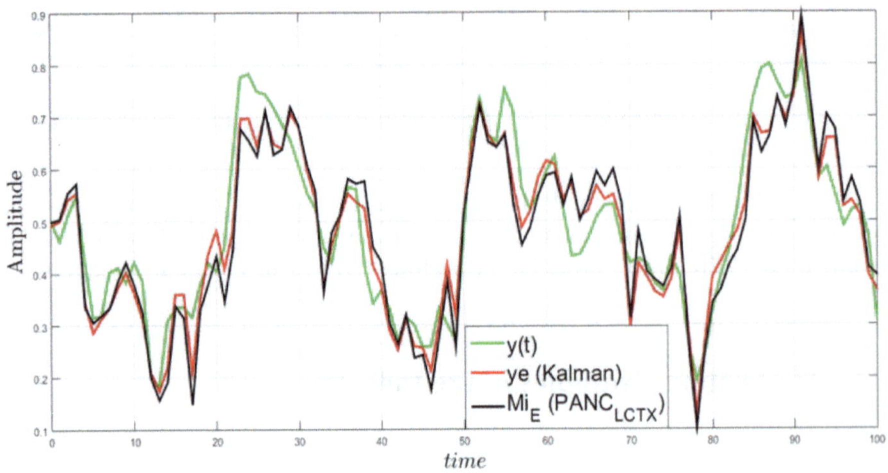

Fig. 5.13 Results of Kalman and $PANC_{LCTX}$ in a Plant of Automation. *Source* Mathworks, Kalman Filter Design (2019)

The Matlab code presented in the reference (Mathworks Inc., 2019) [12] is used to create a measurement of a discrete plant output with noise, yv(t), and the output of Kalman Filter, ye(t). The steady-state Kalman filter design is used, considering the process noise covariance, Q, equal to 2.3 and the sensor noise covariance, R, equal to 1 as the reference.

The $PANC_{LCTX}$ code for one cell is added for comparison purposes, using the Eq. (5.17) presented earlier at the session 3.

Figure 5.13 presents the comparison results.

The input signal must be normalized to the limits of the PAL2v, between 0 and 1, to be handled by the $PANC_{LCTX}$. After it go through the cell, the signal must be denormalized back to the original scale.

The true response (y(t)), the estimated response by Kalman (ye(t)) and yv(t) (from Fig. 5.12) where normalized adding 5 for all values and divided by 10, in order to compare with the output μ_E of the $PANC_{LCTX}$.

After several simulations, as observed by Fig. 5.13, considering F_L equal to 0.643 applied to the $PANC_{LCTX}$ (black line) presents a very close response to the Kalman filter (red line) in the simulated scenario.

5.4.2 *Average Extractor with $PANC_{LCTX}$*

In this scenario, we used 1 to 4 $PANC_{LCTX}$ in cascade (output of one cell is applied to the input of the following cell). Ones and zeros are applied alternately to the input of the first cell.

Figure 5.14 presents the example considering the F_L of 0.25.

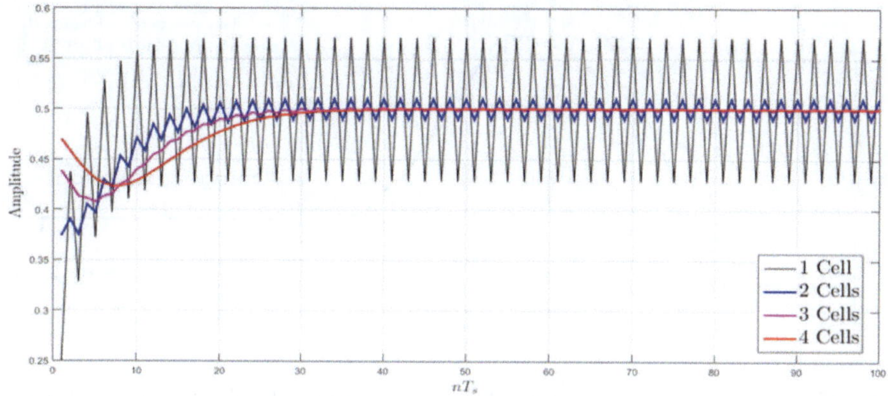

Fig. 5.14 Results of average paraconsistent extraction after alternately 0 and 1 be applied for 1 to 4 $PANC_{LCTX}$

Figure 5.14 shows that all cells present a result around the average of 0.5. As more cells in cascade, more precise is the average extractor.

This arrangement allows the average extraction of a variable with the advantage from the original proposed using $PANC_L$ [10] that the F_L of the $PANC_{LCTX}$ can be adjusted as required by design. Also, because the flexibility of F_L, the number of cells can be optimized, resulting in half the number of cells than for the same scenario.

5.4.3 *Temperature Measurement with $PANC_{LCTX}$*

Considering that the average extraction is a filter for variables with high ripple, we can then use the $PANC_{LCTX}$ structured as the previous configuration for ambient temperature measurements. In order to check that possibility, we measured the environment temperature using a Texas Instruments LM35D sensor, which has an operating range of 0 to 100 °C and has a linear scale factor of + 10 mV/°C. One analog-to-digital converter (ADC) of 10 bits inside the processor is used to convert the voltage from the sensor in a digital number and then convert to the corresponding range of {0,1} used by the paraconsistent logic.

Figure 5.15 presents the temperature measured and the average extractor performing the filtering process. In this configuration two cascaded $PANC_{LCTX}$ with F_L adjustment of 0.25 were used.

The temperature signal has some sparks around the average value (black line). As the temperature is stable the filter extract well the average temperature. Besides that, as a filter, it introduces a response delay to the output. Other values of F_L and the number of cells in the cascade can be used depending on the time constant required.

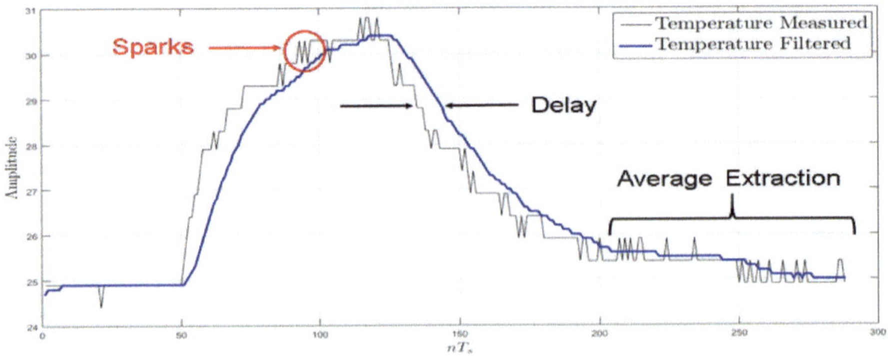

Fig. 5.15 Results of average extraction or filtering process applied to the measured temperature signal

As demonstrated in this study, besides applications in neural networks, the $PANC_{LCTX}$ standalone or in paraconsistent analysis networks can be used as estimator, average extractor or filter expanding the range of applications in the industry.

5.5 Conclusions

Paraconsistent Artificial Neural Cell learning ($PANC_L$) is one of the fundamental cells of a Paraconsistent Artificial Neural Network (PAN_{NET}). This article presented the Paraconsistent Artificial Neural Cell of learning by contradiction extraction ($PANC_{LCTX}$), which has as the main characteristic the ability to learn or unlearn a pattern of input, according to the speed defined by the learning (F_L) and unlearning factors (F_{UL}). The full algorithm of the cell has been presented in order to monitor the variation of the input pattern, which if remain long enough (given numbers of iterations) it will be completely learned by the cell (degree of evidence on the output (μ_E)). This new cell can be applied in any PAN_{NET} that has already been built, just adjusting the learning and the unlearning factors accordingly. The basic equation of $PANC_{LCTX}$ allows, if required, the input pattern to be stored inside the cell. To do this, simply make the F_L and F_{UL} equal to zero after the output μ_E learn the value of the input. This feature can be useful when teaching certain knowledge to a PAN_{NET} and, after full learning, it cannot be influenced by the external environment. As demonstrated in this study, besides applications in neural networks, the $PANCL_{CTX}$ standalone or in paraconsistent analysis networks can be used as an estimator, average extractor, or filter, expanding the range of PAL2v applications in the industry. The results shown in this work indicate that $PANC_{LCTX}$ has a good theoretical foundation and new research will show its application in other areas of engineering and Artificial Intelligence.

Acknowledgments This chapter is dedicated to Prof. Seiki Akama, who has excelled in investigating new techniques for applications of non-classical logics and among them the paraconsistent logic. We wish him well on his 60th birthday.

References

1. Da Costa, N., De Ronde, C.: The paraconsistent logic of quantum superpositions. Found. Phy. **43**(7), 845–858 (2013)
2. Da Costa, N.C.A.: On the theory of inconsistent formal systems. Notre Dame J. Formal Logic (2005)
3. Abe, J.M., Prado, J.C.A., Nakamatsu, K.: Paraconsistent artificial neural network: applicability in computer analysis of speech productions. In: [s.l: s.n.]
4. Da Silva Filho, J.I.: Treatment of uncertainties with algorithms of the paraconsistent annotated logic. J. Intell. Learn. Syst. Appl. **4**(2), 144–153 (2012)
5. Da Silva Filho, J.I., Lambert-Torres, G., Abe, J.M.: Uncertainty treatment using paraconsistent logic: introducing paraconsistent artificial neural networks. Frontiers in Artificial Intelligence and Applications, IOS Press (2010)
6. Coelho, M.S. et al.: Hybrid PI controller constructed with paraconsistent annotated logic. Control Engineering Practice (2019)
7. De Carvalho Jr, A., et al.: A study of paraconsistent artificial neural cell of learning applied as PAL2v Filter. IEEE Lat. Am. Trans. **16**(1), 202–209 (2018)
8. Da Silva Filho, J.I. et al.: Paraconsistent artificial neural network for structuring statistical process control in electrical engineering. In Akama S. (eds) Towards Paraconsistent Engineering. Intelligent Sys Ref Library, Vol. 110, Pp. 77–102, Springer (2016)
9. Abe J.M.: Paraconsistent artificial neural networks: an introduction. In: Negoita M.G., Howlett R.J., Jain L.C. (eds.), Knowledge-Based Intelligent Information and Engineering Systems. KES 2004. Lecture Notes in Computer Science, vol. 3214, pp. 942–948, Springer.
10. Da Silva Filho, J.I.: Algorithms based on paraconsistent annotated logic for applications in expert systems. JM Segura, AC Reiter (eds.), Expert system software: engineering, advantages and applications, Chap. 1. Nova Science Publishers (2011)
11. Lathi, B.P.; Green, R.: Linear systems and signals. Third ed. [s.l.] Oxford University Press (2017)
12. The Mathworks Inc.: Kalman Filter Design. Available at: https://www.mathworks.com/help/control/examples/kalman-filter-design.html, accessed May 2020

Chapter 6
Probabilistic Autoepistemic Equilibrium Logic

Pedro Cabalar, Jorge Fandinno, and Luis Fariñas del Cerro

Contents

Abstract In this short note, we consider the definition of an *Probabilistic Epistemic Logic* (PE) and its non-monotonic extension, we call *Probabilistic Autoepistemic Equilibrium Logic* (PAEE). PE introduces a probabilistic modality that allows expressing lower bounds on conditional probability constructs. Regular (non-probabilistic) modal epistemic operators **K** and **M** can be defined as derived constructs in PE so that, in fact, for that modal epistemic fragment, PE collapses into modal logic KD45. The non-monotonic extension of PE follows the same steps as *Equilibrium Logic* [7], the main logical characterisation of Answer Set Programming [1], ASP. Equilibrium logic consists in a selection among the models of a theory under the intermediate logic of *Here-and-There* (HT) [5]. Similarly, we define the combination of PE with HT, we call PEHT, and then, define a model selection criterion that gives rise to the non-monotonic formalism of Autoepistemic PE. We end up showing that, if we consider again the modal epistemic fragment of the syntax, PAEE

Partially funded by Xunta de Galicia and the European Union, GPC ED431B 2022/33, by the Spanish Ministry of Science and Innovation (grant PID2020-116201GB-I00) and by the National Science Foundation (NSF 95-3101-0060-402).

P. Cabalar
University of Corunna, A Coruña, Spain
e-mail: cabalar@udc.es

J. Fandinno
University of Nebraska at Omaha, Omaha, USA
e-mail: jorgefandinno@unomaha.edu

L. Fariñas del Cerro (✉)
IRIT, University of Toulouse, CNRS, Toulouse, France
e-mail: farinas@irit.fr

J. M. Abe (ed.), *Advances in Applied Logics*, Intelligent Systems Reference Library 243,
https://doi.org/10.1007/978-3-031-35759-6_6

collapses to *Epistemic Specifications* [4], a well-known epistemic extension of ASP and we use a previous result to illustrate how the further addition of the excluded middle axiom, eventually produces *Autoepistemic Logic* [6] as a particular case.

6.1 Syntax and Semantics of PE

The syntax starts from some (countable) set of atoms At we call the propositional *signature*. Formulas φ are defined following the syntax:

$$\varphi ::= p \mid \bot \mid \varphi \wedge \varphi \mid \varphi \vee \varphi \mid \varphi \to \varphi \mid \mathbf{P}(\varphi \mid \varphi) \succeq x$$

where $p \in At$ and x is any constant real number $x \in [0, 1]$. A *theory* Γ is a set of formulas. Intuitively, the reading of the modal operator $\mathbf{P}(\varphi \mid \psi) \succeq x$ is:

the probability of φ conditioned to ψ is at least x.

We will also use the following derived propositional operators $\neg\varphi \stackrel{\text{def}}{=} (\varphi \to \bot)$, $\top \stackrel{\text{def}}{=} \neg\bot$ and $\varphi \leftrightarrow \psi \stackrel{\text{def}}{=} (\varphi \to \psi) \wedge (\psi \to \varphi)$ plus a modal derived operator $\mathbf{P}(\varphi \mid \psi) \succ x$ that is defined as:

$$\mathbf{P}(\varphi \mid \psi) \succ x \quad \stackrel{\text{def}}{=} \quad \neg(\mathbf{P}(\neg\varphi \mid \psi) \succeq 1 - x)$$

and whose intuitive reading is:

the probability of φ conditioned to ψ is greater than x.

We will sometimes use the construct $\mathbf{P}(\varphi)$ as an abbreviation of $\mathbf{P}(\varphi \mid \top)$ regardless of the comparison symbol we use $\succeq$ or $\succ$. In fact, new derived operators using other comparison symbols will be introduced later on. Intuitively, $\mathbf{P}(\varphi \mid \psi) \succeq x$ is a modal necessity operator, whereas $\mathbf{P}(\varphi \mid \psi) \succ x$ is its dual possibility operator.

As always, a *propositional interpretation* I is a set of atoms from the signature, $I \subseteq At$. Each possible interpretation can be seen as a different state of affairs of the real world. To represent the agent's beliefs we will use a probability distribution over these states of affairs. A *(probabilistic) belief view* π is a probability distribution over interpretations $\pi : 2^{At} \to [0, 1]$ so that $\sum_{I \subseteq At} \pi(I) = 1$. The set of *worlds for* π is defined as $W_\pi \stackrel{\text{def}}{=} \{I \subseteq At \mid \pi(I) > 0\}$, so it collects all interpretations that are assigned a strictly positive probability by belief view π, that is, those that effectively are among the agent's beliefs, with some non-zero probability. Note that W_π cannot be empty[1], since the sum of all probabilities for interpretations must be 1.

A *belief interpretation* is a pair (π, I) where π is a belief view and $I \subseteq At$ is an interpretation that accounts for the *real world*. Note that π represents beliefs and not

[1] This holds even for $At = \emptyset$, where the only possible interpretation would be $\emptyset$ and the only possible belief view assigns $\pi(\emptyset) = 1$, so $W_\pi = \{\emptyset\}$.

knowledge: as a result, it may be the case that $I \notin W_\pi$, that is, the agent may believe that the real world I has a probability $\pi(I) = 0$.

Definition 6.1 *(Satisfaction)* We define the satisfaction of formula φ by a belief interpretation (π, I), written $(\pi, I) \models \varphi$, recursively as follows:

1. $(\pi, I) \not\models \bot$
2. $(\pi, I) \models p$ if $p \in I$
3. $(\pi, I) \models \varphi \wedge \psi$ if $(\pi, I) \models \varphi$ and $(\pi, I) \models \psi$
4. $(\pi, I) \models \varphi \vee \psi$ if $(\pi, I) \models \varphi$ or $(\pi, I) \models \psi$
5. $(\pi, I) \models \varphi \rightarrow \psi$ if $\pi, I \not\models \varphi$ or $(\pi, I) \models \psi$
6. $(\pi, I) \models \mathbf{P}(\varphi \mid \psi) \succeq x$ if $\pi(\psi) = 0$ or $\pi(\varphi \wedge \psi)/\pi(\psi) \geq x$

where the application of $\pi(\phi)$ on any formula ϕ we used in the last item simply stands for:

$$\pi(\phi) \stackrel{\text{def}}{=} \sum\{\pi(J) \mid J \subseteq 2^{At}, (\pi, J) \models \phi\}$$ □

According to this definition, it is easy to see that $\pi(\bot) = 0$, $\pi(\neg\varphi) = 1 - \pi(\varphi)$ and $\pi(\top) = 1$. In this context, implication $\varphi \rightarrow \psi$ is classical and amounts to $\neg\varphi \vee \psi$. Similarly, the satisfaction of negation $(\pi, I) \models \neg\varphi$ amounts to $(\pi, I) \not\models \varphi$.

Proposition 6.1 $(\pi, I) \models \mathbf{P}(\varphi \mid \psi) \succ x$ *iff both* $\pi(\psi) \neq 0$ *and* $\pi(\varphi \wedge \psi)/\pi(\psi) > x$.

Proof We start observing:

$$\begin{aligned}
(\pi, I) \models \mathbf{P}(\varphi \mid \psi) \succ x \quad &\Leftrightarrow \quad (\pi, I) \models \neg\mathbf{P}(\neg\varphi \mid \psi) \succeq 1 - x \\
&\Leftrightarrow \quad (\pi, I) \not\models \mathbf{P}(\neg\varphi \mid \psi) \succeq 1 - x \\
&\Leftrightarrow \quad \pi(\psi) \neq 0 \text{ and } \pi(\neg\varphi \wedge \psi)/\pi(\psi) < 1 - x \\
&\Leftrightarrow \quad \pi(\psi) \neq 0 \text{ and } \pi(\neg\varphi \wedge \psi) < (1 - x) \cdot \pi(\psi)
\end{aligned}$$

We will prove that, when $\pi(\psi) \neq 0$, the last conjunct is equivalent to $\pi(\varphi \wedge \psi) > x \cdot \pi(\psi)$. To this aim, note that $\pi(\psi) = \pi(\varphi \wedge \psi) + \pi(\neg\varphi \wedge \psi)$ and so:

$$\begin{aligned}
& \pi(\neg\varphi \wedge \psi) < (1 - x) \cdot \pi(\psi) \\
\Leftrightarrow \quad & \pi(\neg\varphi \wedge \psi) < (1 - x) \cdot \pi(\varphi \wedge \psi) + (1 - x) \cdot \pi(\neg\varphi \wedge \psi) \\
\Leftrightarrow \quad & \pi(\neg\varphi \wedge \psi) - (1 - x) \cdot \pi(\neg\varphi \wedge \psi) < \pi(\varphi \wedge \psi) - x \cdot \pi(\varphi \wedge \psi) \\
\Leftrightarrow \quad & x \cdot \pi(\neg\varphi \wedge \psi) < \pi(\varphi \wedge \psi) - x \cdot \pi(\varphi \wedge \psi) \\
\Leftrightarrow \quad & x \cdot (\pi(\neg\varphi \wedge \psi) + \pi(\varphi \wedge \psi)) < \pi(\varphi \wedge \psi) \\
\Leftrightarrow \quad & x \cdot \pi(\psi) < \pi(\varphi \wedge \psi)
\end{aligned}$$ □

As a result of Proposition 6.1, given that $\pi(\top) = 1$, it is easy to see that:

$$\begin{aligned}
\mathbf{P}(\varphi) \succeq x \;&\Leftrightarrow\; \mathbf{P}(\varphi \mid \top) \succeq x \;\Leftrightarrow\; \pi(\varphi) \geq x \\
\mathbf{P}(\varphi) \succ x \;&\Leftrightarrow\; \mathbf{P}(\varphi \mid \top) \succ x \;\Leftrightarrow\; \pi(\varphi) > x
\end{aligned}$$

A belief interpretation (π, I) is a *belief model* of a theory Γ if $(\pi, J) \models \varphi$ for all $\varphi \in \Gamma$ and $J \in W_\pi \cup \{I\}$. We say that a belief view π is an *epistemic model* of a theory Γ, abbreviated as $\pi \models \Gamma$, when $(\pi, J) \models \varphi$ for all $\varphi \in \Gamma$ and all $J \in W_\pi$. A formula φ is a *tautology* if $(\pi, I) \models \varphi$ for any belief interpretation (π, I). We call *Probabilistic Epistemic Logic* (PE) to the logic induced by all tautologies.

The following are some interesting derived operators and their induced semantics:

$$
\begin{array}{rcll}
\mathbf{K}\varphi & \stackrel{\text{def}}{=} & \mathbf{P}(\varphi) \succeq 1 & \Leftrightarrow (\pi, I) \models \varphi \text{ for all } I \in W_\pi \\
\mathbf{M}\varphi & \stackrel{\text{def}}{=} & \mathbf{P}(\varphi) \succ 0 & \Leftrightarrow (\pi, I) \models \varphi \text{ for some } I \in W_\pi \\
\mathbf{P}(\varphi \mid \psi) \preceq x & \stackrel{\text{def}}{=} & \neg(\mathbf{P}(\varphi \mid \psi) \succ x) & \Leftrightarrow \pi(\psi) = 0 \text{ or } \pi(\varphi \wedge \psi)/\pi(\psi) \leq x \\
\mathbf{P}(\varphi \mid \psi) \prec x & \stackrel{\text{def}}{=} & \neg(\mathbf{P}(\varphi \mid \psi) \succeq x) & \Leftrightarrow \pi(\psi) > 0 \text{ and } \pi(\varphi \wedge \psi)/\pi(\psi) < x \\
\mathbf{P}(\varphi \mid \psi) \hat{=} x & \stackrel{\text{def}}{=} & \mathbf{P}(\varphi \mid \psi) \succeq x & \\
 & & \wedge \mathbf{P}(\varphi \mid \psi) \preceq x & \Leftrightarrow \pi(\psi) = 0 \text{ or } \pi(\varphi \wedge \psi)/\pi(\psi) = x \\
\mathbf{P}(\varphi \mid \psi) = 1 & \stackrel{\text{def}}{=} & \mathbf{P}(\varphi \mid \psi) \succeq 1 & \Leftrightarrow \mathbf{K}(\psi \rightarrow \phi) \\
\mathbf{P}(\varphi \mid \psi) = x & \stackrel{\text{def}}{=} & \mathbf{M}\psi \wedge \mathbf{P}(\varphi \mid \psi) \hat{=} x & \Leftrightarrow \pi(\psi) > 0 \text{ and } \pi(\varphi \wedge \psi)/\pi(\psi) = x \\
 & & & \quad \text{for } x < 1
\end{array}
$$

Notice that $\mathbf{P}(\varphi \mid \psi) \hat{=} x$ is a weak assertion in the sense that it is trivially true when $\pi(\psi) = 0$ (that is, there are no worlds satisfying ψ). The stronger version $\mathbf{P}(\varphi \mid \psi) = x$ depends on the value chosen for x. If $x = 1$ this amounts to checking $\mathbf{P}(\varphi \mid \psi) \succeq 1$ because a probability cannot have a value larger than 1. Note that this formula is equivalent to $\mathbf{K}(\psi \rightarrow \varphi)$, that is, we just check that all worlds satisfying ψ also satisfy φ. When $\pi(\psi) = 0$, there are no worlds in which ψ and the probability of the conditional is trivially $x = 1$. For this reason, when $x < 1$, we must have $\pi(\psi) > 0$ because, as we just said, $\pi(\psi) = 0$ would make the conditional trivially true and require probability $x = 1$. The formula $\mathbf{M}\,\psi$ is used to force $\pi(\psi) > 0$, that together with $\mathbf{P}(\varphi \mid \psi) \hat{=} x$, produces the expected result. We can also observe that:

$$
\begin{array}{rcl}
\mathbf{P}(\varphi \mid \psi) \succeq 1 & \Leftrightarrow & \mathbf{K}(\psi \rightarrow \varphi) \\
\mathbf{P}(\varphi \mid \psi) \succ 0 & \Leftrightarrow & \mathbf{M}(\varphi \wedge \psi)
\end{array}
$$

We say that a formula is *epistemic* when all its modal opearors are of the form **K** or **M**. By a simple inspection of the derived semantics for **K** and **M**, the following result can be easily checked:

Theorem 6.1 *Let φ be an epistemic formula. Then $(\pi, I) \models \varphi$ iff $W_\pi, I \models \varphi$ in modal logic KD45.* □

6.2 Probabilistic Autoepistemic Equilibrium Logic

We define now the combination of PE with the intermediate logic of HT. In the latter, interpretations have the form of pairs $\langle H, T \rangle$ of sets of atoms where H (called the "here" world) is a subset of T (the "there" world). We define an PEHT-*interpretation* as a triple $\langle \pi, H, T \rangle$ where $H \subseteq T \subseteq At$ and π is a belief view. When $H = T$ we say that the interpretation is *total* and we just write it as a pair $\langle \pi, T \rangle$.

Definition 6.2 *(PEHT-satisfaction)* A PEHT-interpretation *satisfies* a formula φ, written $\langle \pi, H, T \rangle \models \varphi$, if the following recursive conditions hold:

- $\langle \pi, H, T \rangle \not\models \bot$
- $\langle \pi, H, T \rangle \models p$ iff $p \in H$
- $\langle \pi, H, T \rangle \models \varphi \wedge \psi$ iff $\langle \pi, H, T \rangle \models \varphi$ and $\langle \pi, H, T \rangle \models \psi$
- $\langle \pi, H, T \rangle \models \varphi \vee \psi$ iff $\langle \pi, H, T \rangle \models \varphi$ or $\langle \pi, H, T \rangle \models \psi$
- $\langle \pi, H, T \rangle \models \varphi \rightarrow \psi$ iff both:
  (i) $\langle \pi, T \rangle \models \varphi \rightarrow \psi$ and
  (ii) $\langle \pi, H, T \rangle \not\models \varphi$ or $\langle \pi, H, T \rangle \models \psi$
- $\langle \pi, H, T \rangle \models \mathbf{P}[\varphi \mid \psi] \succeq x$ if $\pi(\psi) = 0$ or $\pi(\varphi \wedge \psi)/\pi(\psi) \geq x$

where the application of $\pi(\phi)$ on any formula ϕ we used in the last item simply stands for:

$$\pi(\phi) \stackrel{\text{def}}{=} \sum \{\pi(J) \mid J \subseteq 2^{At}, \langle \pi, T \rangle \models \phi\} \qquad \square$$

As usual, we say that $\langle \pi, H, T \rangle$ is a *model* of a theory Γ, in symbols $\langle \pi, H, T \rangle \models \Gamma$, iff $\langle \pi, H, T \rangle \models \varphi$ for all $\varphi \in \Gamma$. We define PEHT-tautologies as formulas satisfied by every PEHT-interpretation, as expected. PEHT is the logic induced by all PEHT-tautologies.

Definition 6.3 *(Equilibrium model)* A set of atoms T is a *π-equilibrium model* of a theory Γ if $\langle \pi, T \rangle \models \Gamma$ and there is no $H \subset T$ s.t $\langle \pi, H, T \rangle \models \Gamma$. $\square$

We denote the set of π-equilibrium models of Γ as $EQ[\pi, \Gamma]$.

Definition 6.4 *(Probabilistic world view)* A belief view π is a *probabilistic world view* for a theory Γ if:

$$W_\pi = EQ[\pi, \Gamma] \qquad \square$$

We define the Probabilistic Autoepistemic Equilibrium Logic (PAEE) as the logic induced by probabilistic world views.

Epistemic Specifications were defined by Gelfond in [4] for an extension of logic programs with epistemic literals in the rule conditions (or bodies). In [2], a straightforward extension for covering the syntax of arbitrary epistemic formulas was provided.

Theorem 6.2 *Let Γ be an epistemic theory and π some belief view. Then π is a probabilistic world view of Γ iff W_π is a world view of Γ in the sense of epistemic specifications as in [2].*

This relation is one-to-many. We may have several π with the same W_π. This also means, for instance, that when we look at the worlds W_π induced by each probabilistic world view π in PAEE and we restrict the syntax to epistemic specifications, we essentially get Gelfond's world views as originally defined in [4].

Moreover, according to Proposition 1 in [2], the world views[2] of an epistemic theory Γ we get from Autoepistemic Logic [6] just correspond to Gelfond's world views for $\Gamma \cup (EM)$ where (EM) stands for the axiom of excluded middle:

$$p \vee \neg p \tag{EM}$$

for every atom $p \in At$. As a consequence, if we consider PAEE plus the (EM) axiom we obtain a *probabilistic proper extension* of Autoepistemic Logic.

6.3 Conclusions

We have presented an expressive non-monotonic formalism, PAEE, whose monotonic basis PEHT constitutes a combination of the logic of Here-and-There plus the well-known modal logic KD45, but further generalised for dealing with probabilities. Similarly, the non-monotonic formalism, PAEE, constitutes a probabilisitc generalisation of Gelfond's epistemic specifications and, when the excluded-middle axiom is added, of Moore's Autoepistemic Logic.

For future work, we plan to investigate the representation of probabilistic independence among atoms or formulas and the incorporation of the principle of indifference (without further information, all states of affairs have equal probability). Also, we plan to investigate the formal relation to the probabilisitc logic programming formalism of ProbLog [8] and to other modal approaches for probabilities such as [3].

References

1. Brewka, G., Eiter, T., Truszczyński, M.: Answer set programming at a glance. Commun. ACM **54**(12), 92–103 (2011)
2. Cabalar, P., Fandinno, J., Fariñas del Cerro, L.: Autoepistemic answer set programming. Artif. Intell. **289**, 103382 (2020). https://doi.org/10.1016/j.artint.2020.103382
3. Fagin, R., Halpern, J.Y., Megiddo, N.: A logic for reasoning about probabilities. Inf. Comput. **87**(1/2), 78–128 (1990). https://doi.org/10.1016/0890-5401(90)90060-U
4. Gelfond, M.: Logic programming and reasoning with incomplete information. Ann. Math. Artif. Intell. **12**(1–2), 89–116 (1994). https://doi.org/10.1007/BF01530762
5. Heyting, A.: Die formalen Regeln der intuitionistischen Logik, pp. 42–56. Sitzungsberichte der Preussischen Akademie der Wissenschaften, Physikalisch-mathematische Klasse pp (1930)
6. Moore, R.C.: Semantical considerations on nonmonotonic logic. Artif. Intell. **25**(1), 75–94 (1985). https://doi.org/10.1016/0004-3702(85)90042-6

[2] Actually called *theory expansions* in the original terminology [6].

7. Pearce, D.: A new logical characterisation of stable models and answer sets. In: NMELP. Lecture Notes in Computer Science, vol. 1216, pp. 57–70. Springer (1997)
8. Raedt, L.D., Kimmig, A., Toivonen, H.: Problog: a probabilistic prolog and its application in link discovery. In: Veloso, M.M. (ed.) IJCAI 2007, Proceedings of the 20th International Joint Conference on Artificial Intelligence, Hyderabad, India, January 6–12, 2007, pp. 2462–2467 (2007). http://ijcai.org/Proceedings/07/Papers/396.pdf

Chapter 7
Rough-Set-Base Data Analysis: Theoretical Basis and Applications

Yasuo Kudo and Tetsuya Murai

Contents

Abstract Rough set theory, originally proposed by Z. Pawlak, provides a mathematical basis of set-based approximation of concepts and logical data analysis. In this chapter, we review an approach of rough-set-based data analysis by the authors.

Y. Kudo (✉)
College of Information and Systems, Muroran Institute of Technology, Hokkaido, Japan
e-mail: kudo@csse.muroran-it.ac.jp

T. Murai
Department of Information Systems Engineering, Chitose Institute of Science and Technology, Hokkaido, Japan
e-mail: t-murai@photon.chitose.ac.jp

J. M. Abe (ed.), *Advances in Applied Logics*, Intelligent Systems Reference Library 243,
https://doi.org/10.1007/978-3-031-35759-6_7

7.1 Introduction

Rough set theory, originally proposed by Pawlak [21, 22], provides a mathematical basis of set-based approximation of concepts and logical data analysis. In this chapter, we review an approach of rough-set-based data analysis by the authors. This approach is mainly based on (1) reduction of decision table [13], and (2) evaluation of relative reduct by roughness of partition [12].

The rest of this chapter is constructed as follows: Sect. 7.2 introduces basic concepts of rough sets for data analysis; in particular, decision table, relative reduct, discernibility matrix, and decision rule. In Sect. 7.3, we review an approach to generate as many relative reducts as possible from decision tables with numerous condition attributes [13]. This approach is based on reduction of decision table and iteration of extracting as many relative reducts as possible from reduced decision tables. In Sect. 7.4, we review an evaluation method of relative reducts by roughness of partitions obtained from relative reducts [12]. Section 7.5 introduces an example of rough-set-based data analysis based on this approach [16]. This approach is applied to DNA data analysis. Finally, Sect. 7.6 summarizes this chapter.

7.2 Rough Sets

In this section, we review the rough set theory, in particular, decision table, relative reduct, and discernibility matrix. Note that this section is based on [19, 23].

7.2.1 Decision Table and Lower and Upper Approximations

Generally, data analysis subjects by rough sets are described by decision tables. Formally, a *decision table* is characterized by the following triple:

$$DT = (U, C, d), \tag{7.1}$$

where U is a finite and nonempty set of *objects*, C is a finite and nonempty set of *condition attributes*, and d is a *decision attribute* such that $d \notin C$. Each attribute $a \in C \cup \{d\}$ is a function $a : U \to V$, where V is a set of values of attributes.

Indiscernibility relations based on subsets of attributes provide classifications of objects in decision tables. For any set of attributes $A \subseteq C \cup \{d\}$, the *indiscernibility relation* R_A is the following binary relation on U:

$$R_A = \{(x, y) \mid a(x) = a(y), \forall a \in A\}. \tag{7.2}$$

If a pair (x, y) is in R_A, then two objects x and y are indiscernible with respect to all attributes in A. It is well-known that any indiscernibility relation is an equivalence relation and equivalence classes by an equivalence relation consist of a *partition* on the domain of the equivalence relation. In particular, the indiscernibility relation R_d based on the decision attribute d provides a partition $\mathcal{D} = \{D_1, \ldots, D_k\}$, and each element $D_i \in \mathcal{D}$ is called a *decision class*.

Classifying objects with respect to condition attributes provides approximation of decision classes. Formally, for any set $B \subseteq C$ of condition attributes and any decision class $D_i \in \mathcal{D}$, we let:

$$\underline{B}(D_i) = \{x \in U \mid [x]_B \subseteq D_i\}, \tag{7.3}$$
$$\overline{B}(D_i) = \{x \in U \mid [x]_B \cap D_i \neq \emptyset\}, \tag{7.4}$$

where the set $[x]_B$ is the *equivalence class* of x by the indiscernibility relation R_B. The set $\underline{B}(D_i)$ and the set $\overline{B}(D_i)$ are called *lower approximation* and *upper approximation* of the decision class D_i with respect to B, respectively. Note that the lower approximation $\underline{B}(D_i)$ is the set of objects that are correctly classified to the decision class D_i by checking all attributes in B. A decision table is called *consistent* if and only if $\underline{C}(D_i) = D_i = \overline{C}(D_i)$ holds for all decision classes $D_i \in \mathcal{D}$.

Example 7.1 Table 7.1 is an example of a decision table we use in this chapter, and consists of the following objects: $U = \{x_1, \ldots, x_7\}$, $C = \{c_1, \ldots, c_8\}$, and d. The decision attribute d provides the following three decision classes: $D_1 = \{x_1, x_5\}$, $D_2 = \{x_3, x_4, x_7\}$ and $D_3 = \{x_2, x_6\}$.

For any subset $B \subseteq C$ of condition attributes and any decisoin class $D_i \in \mathcal{D}$, classification ability of B for classifying objects into correct decision classe D_i are evaluated by *accuracy* $\alpha_B(D_i)$ and *quality of approximation* $\gamma_B(D_i)$, respectively:

$$\alpha_B(D_i) = \frac{|\underline{B}(D_i)|}{|\overline{B}(D_i)|}, \tag{7.5}$$
$$\gamma_B(D_i) = \frac{|\underline{B}(D_i)|}{|D_i|}, \tag{7.6}$$

Table 7.1 Decision table

U	c_1	c_2	c_3	c_4	c_5	c_6	c_7	c_8	d
x_1	1	1	1	1	1	1	1	2	1
x_2	3	1	3	2	2	1	1	2	3
x_3	2	3	2	1	2	2	1	1	2
x_4	2	2	2	1	2	2	3	1	2
x_5	2	2	3	1	1	1	1	2	1
x_6	3	2	1	2	2	1	2	2	3
x_7	1	1	1	1	1	1	1	1	2

where $|X|$ means the cardinality of the subset $X \subseteq U$. The accuracy $\alpha_B(D_i)$ represents how the objects in D_i are approximated by information of B. The quality of approximation $\gamma_B(D_i)$ represents how the objects in D_i are correctly classified into D_i by information of B. For the partition $\mathcal{D} = \{D_1, \ldots, D_k\}$ by decision classes, the *quality of approximation* $\gamma_B(\mathcal{D})$ by B is defined as follows:

$$\gamma_B(\mathcal{D}) = \frac{\sum_{i=1}^{k} |\underline{B}(D_i)|}{|U|}. \tag{7.7}$$

Note that the decision table is consistent if and only if $\gamma_C(\mathcal{D}) = 1$ holds.

7.2.2 Relative Reduct

By checking values of all condition attributes, we can classify all discernible objects of the given decision table to the corresponding decision classes. However, not all condition attributes may need to be checked in the sense that some condition attributes are essential to classify and the other attributes are redundant. A minimal set of condition attributes to classify all discernible objects to correct decision classes is called a relative reduct of the decision table.

For any subset $X \subseteq C$ of condition attributes in a decision table DT, we let:

$$POS_X(\mathcal{D}) = \bigcup_{D_i \in \mathcal{D}} \underline{X}(D_i). \tag{7.8}$$

The set $POS_X(\mathcal{D})$ is called the *positive region* of $\mathcal{D}$ by X. All objects $x \in POS_X(\mathcal{D})$ are classified to correct decision classes by checking all attributes in X. In particular, the set $POS_C(\mathcal{D})$ is the set of all discernible objects in DT.

Here, we define relative reducts formally. A set $A \subseteq C$ is called a *relative reduct* of the decision table DT if the set A satisfies the following conditions:

1. $POS_A(\mathcal{D}) = POS_C(\mathcal{D})$,
2. $POS_B(\mathcal{D}) \neq POS_C(\mathcal{D})$ for any proper subset $B \subset A$.

Note that, in general, there are plural relative reducts in a decision table. The common part of all relative reducts is called the *core* of the decision table.

For example, there are the following four relative reducts in Table 7.1: $\{c_1, c_8\}$, $\{c_4, c_8\}$, $\{c_5, c_8\}$, and $\{c_2, c_3, c_8\}$. The condition attribute c_8 appears in all relative reducts in Table 7.1, and therefore, the core of Table 7.1 is $\{c_8\}$.

7.2.3 Discernibility Matrix

The discernibility matrix is one of the most popular methods to compute all relative reducts in the decision table. Let DT be a decision table with $|U|$ objects. The *discernibility matrix* DM of DT is a $|U| \times |U|$ matrix whose element at i-th row and j-th column is the following set of condition attributes to discern between two objects x_i and x_j:

$$\delta_{ij} = \begin{cases} \{a \in C \mid a(x_i) \neq a(x_j)\}, & \text{if } d(x_i) \neq d(x_j) \text{ and } \{x_i, x_j\} \cap POS_C(\mathcal{D}) \neq \emptyset, \\ \emptyset, & \text{otherwise.} \end{cases} \tag{7.9}$$

Each element $a \in \delta_{ij}$ represents that x_i and x_j are discernible by checking the value of a, and at least one of x_i and x_j is classifiable to its correct decision class by checking all condition attributes. By this definition, it is obvious that 1) $\delta_{ii} = \emptyset$ for all $1 \leq i \leq |U|$ and 2) $\delta_{ij} = \delta_{ji}$ for any $1 \leq i \leq j \leq |U|$ hold. Therefore, the discernibility matrix is a symmetric matrix and in this chapter, we omit the upper triangle part of the discernibility matrix.

Using the discernibility matrix, we get all relative reducts of the decision table as follows:

1. Construct the following logical formula $L(\delta_{ij})$ from each nonempty set $\delta_{ij} = \{a_{k1}, \ldots, a_{kl}\}$ ($i > j$ and $l \geq 1$) in the discernibility matrix:

$$L(\delta_{ij}) : a_{k1} \vee \cdots \vee a_{kl}. \tag{7.10}$$

2. Construct a conjunctive normal form $\bigwedge_{i>j} L(\delta_{ij})$.
3. Transform the conjunctive normal form to the minimal disjunctive normal form:

$$\bigwedge_{i>j} L(\delta_{ij}) \equiv \bigvee_{p=1}^{s} \bigwedge_{q=1}^{t_p} a_{pq}. \tag{7.11}$$

4. For each conjunction $a_{p1} \wedge \cdots \wedge a_{pt_p}$ ($1 \leq p \leq s$) in the minimal disjunctive normal form, construct a relative reduct $\{a_{p1}, \ldots, a_{pt_p}\}$.

Example 7.2 Table 7.2 describes the discernibility matrix of the decision table by Table 7.1. Each nonempty set that appears in the matrix represents the set of condition attributes that we should check to discern the corresponding objects. For example, the set $\delta_{21} = \{c_1, c_3, c_4, c_5\}$ represents that we can distinguish between the objects x_2 and x_1 by comparing values of these objects of at least one of the condition attributes c_1, c_3, c_4, and c_5 in δ_{21}. Note that we omit upper triangular components of the discernibility matrix in Table 7.2. We construct a conjunctive normal form by connecting logical formulas based on nonempty elements in Table 7.2 by (7.10) and (7.11), and transform the conjunctive normal form to the minimal disjunctive normal form as follows:

Table 7.2 The discernibility matrix of Table 7.1

	x_1	x_2	x_3	x_4	x_5	x_6	x_7
x_1	$\emptyset$						
x_2	$\{c_1, c_3, c_4, c_5\}$	$\emptyset$					
x_3	$\{c_1, c_2, c_3, c_5, c_6, c_8\}$	$\{c_1, c_2, c_3, c_4, c_6, c_8\}$	$\emptyset$				
x_4	$\{c_1, c_2, c_3, c_5, c_6, c_7, c_8\}$	$\{c_1, c_2, c_3, c_4, c_6, c_7, c_8\}$	$\emptyset$	$\emptyset$			
x_5	$\emptyset$	$\{c_1, c_2, c_4, c_5\}$	$\{c_2, c_3, c_5, c_6, c_8\}$	$\{c_3, c_5, c_6, c_7, c_8\}$	$\emptyset$		
x_6	$\{c_1, c_2, c_4, c_5, c_7\}$	$\emptyset$	$\{c_1, c_2, c_3, c_4, c_6, c_7, c_8\}$	$\{c_1, c_3, c_4, c_6, c_7, c_8\}$	$\{c_1, c_3, c_4, c_5, c_7\}$	$\emptyset$	
x_7	$\{c_8\}$	$\{c_1, c_3, c_4, c_5, c_8\}$	$\emptyset$	$\emptyset$	$\{c_1, c_2, c_3, c_8\}$	$\{c_1, c_2, c_4, c_5, c_7, c_8\}$	$\emptyset$

$$
\begin{aligned}
&(c_1 \vee c_3 \vee c_4 \vee c_5) \wedge (c_1 \vee c_2 \vee c_3 \vee c_5 \vee c_6 \vee c_8) \\
&\wedge \cdots \wedge (c_1 \vee c_2 \vee c_4 \vee c_5 \vee c_7 \vee c_8) \\
&\equiv (c_1 \wedge c_8) \vee (c_4 \wedge c_8) \vee (c_5 \wedge c_8) \vee (c_2 \wedge c_3 \wedge c_8).
\end{aligned}
$$

Consequently, from this minimal disjunctive normal form, we have the four relative reducts $\{c_1, c_8\}$, $\{c_4, c_8\}$, $\{c_5, c_8\}$, and $\{c_2, c_3, c_8\}$.

7.2.4 Decision Rule

Suppose $a \in C \cup \{d\}$ is any attribute in the decision table and $x \in U$ is any object. We denote the expression $[a = a(x)]$ and call this expression the *descriptor*. The meaning of the descriptor is "the value of the object x at the attribute a is $a(x)$." Descriptors correspond to atomic propositions in propositional logic. Hence, using logical connectivities $\wedge$ (and), $\vee$ (or), $\rightarrow$ (implication), and $\neg$ (not), we can consider logical data analysis based on rough set theory. In particular, we use the following implication obtained from any non-empty subset of condition attributes $B \subseteq C$ and any object $x \in U$ and call the implication the *decision rule* by B and x:

$$
\bigwedge_{a \in B} [a = a(x)] \rightarrow [d = d(x)]. \tag{7.12}
$$

The decision rule obtained from $B = \{a_1, \ldots, a_k\}$ and $x \in U$ means "if the value of x at a_1 is $a_1(x)$ and $\cdots$ and the value of x at a_k is $a_k(x)$, then the value of the decision attribute d is $d(x)$." Hereafter, without any confusion, we denote the antecedent $\bigwedge_{a \in B}[a = a(x)]$ obtained from B and x by (B, x), the conclusion by d and x as (d, x), and the decision rule obtained from B and x as $(B, x) \to (d, x)$, respectively.

Evaluation of decision rule is very important for considering rough-set-based data analysis. Let $(B, x) \to (d, x)$ be any decision rule obtained from $B \subseteq C$ and $x \in U$. The following three criteria, called *certainty*, *coverage*, and *support*, are used for evaluating the decision rules:

$$Cer((B, x) \to (d, x)) = \frac{|[x]_B \cap D_i|}{|[x]_B|}, \tag{7.13}$$

$$Cov((B, x) \to (d, x)) = \frac{|[x]_B \cap D_i|}{|D_i|}, \tag{7.14}$$

$$Supp((B, x) \to (d, x)) = \frac{|[x]_B \cap D_i|}{|U|}, \tag{7.15}$$

where D_i is the decision class that satisfies $x \in D_i$. For any decision rule $(B, x) \to (d, x)$, certainty, coverage, and support of this rule represent evaluation scores of correctness, representation ability, and generality of this rule, respectively.

Construction of decision rules is closely related to extraction of relative reducts. Example 7.3 illustrates importance of selection of relative reducts for generating useful decision rules.

Example 7.3 We use a relative reduct $B = \{c_1, c_8\}$ to construct decision rules $(B, x) \to (d, x)$ from Table 7.1. From the object $x_1 \in U$ and B, we obtain an antecedent $[c_1 = 1] \wedge [c_8 = 2]$. Note that objects in the same equivalence class generate the same antecedent; because $x_6 \in [x_2]_B$, the antecedent $[c_1 = 3] \wedge [c_8 = 2]$ obtained from x_6 is identical to the one obtained from x_2. From the seven objects $x_1, \ldots, x_7 \in U$ and the relative reduct $\{c_1, c_8\}$, the following five decision rules are obtained:

1. $[c_1 = 1] \wedge [c_8 = 2] \to [d = 1]$, $Cer = 1$, $Cov = \frac{1}{2}$, $Supp = \frac{1}{7}$,
2. $[c_1 = 3] \wedge [c_8 = 2] \to [d = 3]$, $Cer = 1$, $Cov = 1$, $Supp = \frac{2}{7}$,
3. $[c_1 = 2] \wedge [c_8 = 1] \to [d = 2]$, $Cer = 1$, $Cov = \frac{2}{3}$, $Supp = \frac{2}{7}$,
4. $[c_1 = 2] \wedge [c_8 = 2] \to [d = 1]$, $Cer = 1$, $Cov = \frac{1}{2}$, $Supp = \frac{1}{7}$,
5. $[c_1 = 1] \wedge [c_8 = 1] \to [d = 2]$, $Cer = 1$, $Cov = \frac{1}{3}$, $Supp = \frac{1}{7}$,

where Cer, Cov, and $Supp$ means certainty, coverage, and support of each decision rule, respectively. These decisoin rules describe some "knowledge" about decision classes. For example, the rule No. 2 $[c_1 = 3] \wedge [c_8 = 2] \to [d = 3]$ represents an important feature of the decision class with $d = 3$; every object with $d = 3$ has $[c_1 = 3]$ and $[c_8 = 2]$. This is based on $Cov = 1$ of this rule.

On the other hand, if we use other relative reduct $\{c_2, c_3, c_8\}$, seven decision rules (details are omitted) are obtained from seven objects. This means that these rules are merely the description of seven cases separately and it is difficult to obtain some "knowledge" about decision classes.

7.3 Heuristic Algorithm for Attribute Reduction Using Reduced Decision Tables

In this section, we review a heuristic approach to generate as many relative reducts as possible from decision tables with numerous condition attributes. This section is based on our manuscript [13].

First, a concept of *reduced decision table* of the given decision table is introduced.

Definition 7.1 Let $DT = (U, C, d)$ be a decision table. A reduced decision table of DT is the following triple:

$$RDT = (U, C', d), \tag{7.16}$$

where U and d are identical to DT. The set of condition attributes C' satisfies the following conditions:

1. $C' \subseteq C$.
2. For any objects x_i and x_j that belong to different decision classes, if x_i and x_j are discernible by R_C, x_i and x_j are also discernible by $R_{C'}$.

The reduced decision table preserves discernibility of objects in the given decision table. In general, there are plural reduced decision tables of the given decision table.

Next, we introduce an algorithm to construct a reduced decision table. It is easy to confirm that Algorithm 1 generates a reduced decision table of the given decision table. In Algorithm 1, we select condition attributes from C at random based on the parameter of base size b, and supply some attributes in elements of the discernibility matrix to preserve discernibility of objects in the given decision table. For generating as many relative reducts as possible from a given decision table with numerous condition attributes, we need to avoid generating the same reduced decision table as long as possible when we use Algorithm 1 repeatedly. Thus, randomness in selecting condition attributes is important to keep variety of reduced decision tables.

Note that the base size b decides the minimum number of condition attributes of the reduced decision table and we need to set b appropriately. If b is too large, it is obvious that there is no merit of using Algorithm 1 at all because the reduced decision table constructed by Algorithm 1 is almost the same (or identical in the case of $b = |C|$) to the original decision table. On the other hand, if b is too small, almost attributes in the reduced decision table are supplied at steps 3–8 in Algorithm 1 to

Algorithm 1 dtr: Decision Table Reduction Algorithm

Require: decision table $DT = (U, C, d)$,
discernibility matrix DM of DT,
base size b
Ensure: reduced decision table (U, C', d)
1: Select b attributes $a_1, \ldots, a_b$ from C at random by sampling without replacement
2: $C' = \{a_1, \ldots, a_b\}$
3: **for all** $\delta_{ij} \in DM$ such that $i > j$ **do**
4: **if** $\delta_{ij} \neq \emptyset$ **and** $\delta_{ij} \cap C' = \emptyset$ **then**
5: Select $c \in \delta_{ij}$ at random
6: $C' = C' \cup \{c\}$
7: **end if**
8: **end for**
9: **return** (U, C', d)

preserve discernibility of objects, which is almost the same to generate a candidate of relative reduct by random selection and it is not our intention.

The following theorem is essential for the attribute reduction algorithm that we propose later.

Theorem 7.1 ([13]) *Let $DT = (U, C, d)$ be a decision table and $RDT = (U, C', d)$ be a reduced decision table of DT constructed by Algorithm 1. A set of condition attributes $A \subseteq C'$ is a relative reduct of RDT if and only if A is a relative reduct of DT.*

Proof Because the set of objects U and the decision attribute d of DT and RDT are identical each other by Definition 7.1, it is obvious that decision classes of both DT and RDT are also identical. Thus, by the definition of relative reducts, it is sufficient to prove that $POS_{C'}(\mathcal{D}) = POS_C(\mathcal{D})$ holds. First, we show $POS_{C'}(\mathcal{D}) \subseteq POS_C(\mathcal{D})$ holds. Let $x \in POS_{C'}(\mathcal{D})$. Then, there exists a decision class D_i such that $[x]_{C'} \subseteq D_i$ holds. The condition (1) in Definition 7.1 implies $R_C \subseteq R_{C'}$, and therefore $[x]_C \subseteq [x]_{C'}$ holds, which concludes $x \in POS_C(\mathcal{D})$. Conversely, we show $POS_C(\mathcal{D}) \subseteq POS_{C'}(\mathcal{D})$ holds. Suppose that $x \notin POS_{C'}(\mathcal{D})$. Then, there exists an object y such that x and y belong to different decision classes and $y \in [x]_{C'}$ holds, i.e., $(x, y) \in R_{C'}$. By the contraposition of the condition (2) in Definition 7.1, $(x, y) \in R_{C'}$ implies $(x, y) \in R_C$, and therefore, $y \in [x]_C$ holds. This implies $x \notin POS_C(\mathcal{D})$, which concludes the proof of the theorem. □

As an example, we show how Algorithm 1 works to generate a reduced decision table from Table 7.1.

Example 7.4 Using Algorithm 1, we construct a reduced decision table RDT of the decision table $DT = (U, C, d)$ described by Table 7.1. Let DM be the discernibility matrix of DT by Table 7.2, and $b = 2$ be the base size as an input of Algorithm 1. At steps 1 and 2 of Algorithm 1, suppose two condition attributes c_1 and c_4 are selected with respect to $b = 2$, and let $C' = \{c_1, c_4\}$. At step 4, because $\delta_{71} = \{c_8\} \neq \emptyset$ and $C' \cap \delta_{71} = \emptyset$ hold, C' is updated to $C' = \{c_1, c_4, c_8\}$ at step 5.

Table 7.3 A reduced decision table of Table 7.1

U	c_1	c_4	c_8	d
x_1	1	1	2	1
x_2	3	2	2	3
x_3	2	1	1	2
x_4	2	1	1	2
x_5	2	1	2	1
x_6	3	2	2	3
x_7	1	1	1	2

Finally, Algorithm 1 generates a reduced decision table $RDT = (U, C', d)$ with $C' = \{c_1, c_4, c_8\}$ described by Table 7.3. RDT has the following two relative reducts: $\{c_1, c_8\}$ and $\{c_4, c_8\}$. As we described in Example 7.2, these are also the relative reducts of Table 7.1.

Thus, we can generate relative reducts of the given decision table DT by generating relative reducts of a reduced decision table RDT constructed from DT by Algorithm 1. If the number of condition attributes in RDT is sufficiently small, we can generate all relative reducts of RDT by exhaustive attribute reduction like the discernibility matrix. Thus, even though the number of condition attributes of the given decision table is numerous, generating reduced decision tables repeatedly and switching the methods of attribute reduction based on the size of each reduced decision table, we can generate many relative reducts (including candidates of relative reduct) instead of applying some heuristic attribute reduction directly to the original decision table.

Using Algorithm 1, we propose the following algorithm of attribute reduction based on generating reduced decision tables and switching exhaustive attribute reduction and heuristic attribute reduction.

Algorithm 2 switches exhaustive attribute reduction and heuristic attribute reduction by the size of decision tables. In Algorithm 2, the size limit L is the threshold for switching attribute reduction methods and if the number of condition attributes of a decision table is smaller than L, Algorithm 2 tries to generate the set of all relative reducts of the decision table. Thus, we need to set the threshold L appropriately. If the number of condition attributes of the given decision table DT is greater than the threshold L, Algorithm 2 repeats I times of generating a reduced decision table RDT and attribute reduction from RDT by selecting the exhaustive method or the heuristic method, and generate the set RED of relative reducts. Note that RED may contain some output with redundancy if the result of the heuristic attribute reduction is not guaranteed to generate relative reducts.

Algorithm 2 Exhaustive/Heuristic Attribute Reduction

Require: decision table $DT = (U, C, d)$,
base size b, size limit L, number of iteration I
Ensure: set of candidates of relative reduct RED

```
RED = ∅
DM ← the discernibility matrix of DT
if |C| ≤ L then
  RED ← result of exhaustive attribute reduction from DT
else
  for i = 1 to I do
    RDT = dtr(DT, DM, b)
    if |C'| ≤ L then
      S ← result of exhaustive attribute reduction from RDT
    else
      S ← result of heuristic attribute reduction from RDT
    end if
    RED = RED ∪ S
  end for
end if
return RED
```

7.3.1 *Conclusion of Section 7.3*

In this section, we reviewed an attribute reduction algorithm to compute as many relative reducts as possible from a decision table with numerous condition attributes. This algorithm is based on generating many reduced decision tables that preserve discernibility of objects in the given decision table and switching exhaustive attribute reduction and heuristic attribute reduction by the number of condition attributes in decision tables. Future issues include adaptive setting of parameters relative to datasets, refinement of the switching method of exhaustive attribute reduction and heuristic attribute reduction, parallelization of attribute reduction using the proposed algorithm and an increased number of experiments using extensive datasets with more numerous condition attributes.

7.4 Evaluation of Relative Reducts Using Partitions

In this section, we review an evaluation method of relative reducts for rough-set-based data analysis. This section is based on our paper [12].

7.4.1 Roughness of Partition and Average of Coverage of Decision Rules

In this section, an evaluation method for relative reducts using partitions constructed from them is introduced. To summarize relative reduct evaluation, we consider that rougher partitions constructed by a relative reduct lead to better evaluation of the relative reduct. However, the quality of partition defined by (7.7) does not consider roughness of partition, although it considers correctness of approximation. In fact, all relative reducts of a consistent decision table, i.e., a decision table $DT = (U, C, d)$ that satisfies $POS_C(\mathcal{D}) = U$, provide crisp approximations of all decision classes, i.e., $\underline{A}(D_i) = D_i = \overline{A}(D_i)$ for any relative reduct A and any decision class D_i, even though the difference of roughness of partitions based on relative reducts. For example, because the decision table illustrated by Table 7.1 is consistent, all relative reducts of Table 7.1 provide crisp approximation, and it is easy to confirm that the score of quality of partition by any relative reduct in Table 7.1 is equal to 1. However, the roughness of partitions based on the relative reducts differs as follows:

- Partition obtained from the relative reduct $\{c_1, c_8\}$: $\{\{x_1\}, \{x_2, x_6\}, \{x_3, x_4\}, \{x_5\}, \{x_7\}\}$,
- Partition obtained from $\{c_4, c_8\}$: $\{\{x_1, x_5\}, \{x_2\}, \{x_3, x_4, x_7\}, \{x_6\}\}$,
- Partition obtained from $\{c_5, c_8\}$: $\{\{x_1, x_5\}, \{x_2\}, \{x_3, x_4\}, \{x_6\}, \{x_7\}\}$,
- Partition obtained from $\{c_2, c_3, c_8\}$: $\{\{x_1\}, \{x_2\}, \{x_3\}, \{x_4\}, \{x_5\}, \{x_6\}, \{x_7\}\}$.

In particular, all equivalence classes in the partition by $\{c2, c3, c8\}$ are singletons. Thus, the quality of approximation is not suitable for evaluating roughness of partition.

In addition, from the viewpoint of rule generation, such rough partitions constructed from relative reducts provide decision rules with higher scores of coverage than those of coverage of decision rules based on fine partitions. Moreover, the correctness of partitions based on relative reducts is guaranteed, because each relative reduct provides the same approximation as the one based on the set of all condition attributes. Thus, we consider evaluating relative reducts using the coverage of decision rules constructed from them.

Here, we consider deriving some relationship between roughness of partitions based on relative reducts and coverage of decision rules based on them. Suppose we fix a non-empty subset of condition attributes $B \subseteq C$. For any equivalence class $[x]_B \in U/R_B$, we define the set

$$Dec_B([x]_B) \stackrel{\text{def}}{=} \{D_i \in \mathcal{D} \mid [x]_B \cap D_i \neq \emptyset\}. \tag{7.17}$$

The set $Dec_B([x]_B)$ corresponds to the set of conclusion(s) of decision rule(s) with the formula (B, x) as the antecedent. Thus, the value defined by

$$N_B \stackrel{\text{def}}{=} \sum_{[x]_B \in U/R_B} |Dec_B([x]_B)| \tag{7.18}$$

is the sum of number of all decision rules constructed from B. Note that the following important property of the set $Dec_B([x]_B)$ is easily obtained:

$$x \in \underline{B}(D_i) \text{ if and only if } Dec_B([x]_B) = \{D_i\}. \tag{7.19}$$

Similarly, we define the set

$$Cond_B(D_i) \stackrel{\text{def}}{=} \{[x]_B \in U/R_B \mid [x]_B \cap D_i \neq \emptyset\}. \tag{7.20}$$

The set $Cond_B(D_i)$ corresponds to the set of antecedents of decision rules with the formula (d, y) for some $y \in D_i$ as conclusion.

From the definitions of the sets $Dec_B([x]_B)$ and $Cond_B(D_i)$, the following relationship is obvious;

$$D_i \in Dec_B([x]_B) \text{ if and only if } [x]_B \in Cond_B(D_i). \tag{7.21}$$

Moreover, the followings equations are easily obtained.

$$\sum_{D_i \in Dec_B([x]_B)} \frac{|[x]_B \cap D_i|}{|[x]|_B|} = 1, \tag{7.22}$$

$$\sum_{[x]_B \in Cond_B(D_i)} \frac{|[x]_B \cap D_i|}{|D_i|} = 1. \tag{7.23}$$

Proofs of these equations are in [12].

These equations represent useful properties for considering the average certainty and the average coverage of decision rules constructed from non-empty subset of condition attributes.

Theorem 7.2 ([12]) *For any non-empty subset $B \subseteq C$ of condition attributes, the average of certainty $ACer(B)$ of all decision rules $(B, x) \to (d, x)$ $(\forall x \in U)$ constructed from equivalence classes $[x]_B \in U/R_B$ and decision classes $[x]_d \in \mathcal{D}$ is calculated by*

$$ACer(B) = \frac{|U/R_B|}{N_B}. \tag{7.24}$$

Similarly, the average of coverage $ACov(B)$ is calculated by

$$ACov(B) = \frac{|\mathcal{D}|}{N_B}. \tag{7.25}$$

Proof We derive the Eq. (7.25). Because $\mathcal{D} = \{D_1, \ldots, D_k\}$ is a partition on U, we can consider the sum of the coverage of all decision rules constructed from B by treating all decision classes $D_i \in \mathcal{D}$ in (7.23). Thus, using the number N_B of all decision rules by (7.18), we simplify the average of coverage of all decision rules constructed from B as

$$
\begin{aligned}
ACov(B) &= \frac{1}{N_B} \sum_{D_i \in \mathcal{D}} \left(\sum_{[x]_B \in Cond_B(D_i)} \frac{|[x]_B \cap D_i|}{|D_i|} \right) \\
&= \frac{1}{N_B} \sum_{D_i \in \mathcal{D}} 1 \quad \text{(by Eq. (7.23))} \\
&= \frac{|\mathcal{D}|}{N_B}.
\end{aligned}
$$

This concludes the derivation of the Eq. (7.25).

Equation (7.24) is also similarly derived; we omit in this proof. □

Theorem 7.2 demonstrates that we can calculate the average certainty and the average coverage of decision rules constructed from non-empty set $B \subseteq C$ by the following three parameters; the number of equivalence classes based on B, the number of decision classes, and the number of decision rules constructed from B. In particular, because the number of decision classes is uniquely determined for any decision table, the average coverage depends only on the number of decision rules.

Moreover, from Theorem 7.2, we can derive the relationship between the roughness of partitions based on relative reducts and the coverage of decision rules based on relative reducts.

Theorem 7.3 ([12]) *Let E and F be relative reducts of a given decision table. The following properties are satisfied:*

1. *If $R_E \subset R_F$ holds, i.e., the partition U/R_F is rougher than the partition U/R_E, then $ACov(E) \leq ACov(F)$ holds.*
2. *If the given decision table is consistent and $R_E \subset R_F$ holds, then $ACov(E) < ACov(F)$ holds.*

Proof 1. By Theorem 7.2, it is sufficient to show that $R_E \subset R_F$ implies $N_E \geq N_F$. It is easy to confirm that the condition $R_E \subset R_F$ implies $[x]_E \subseteq [x]_F$ for any $x \in U$, and there is at least one object $y \in U$ with $[y]_F = \bigcup_{j=1}^{p} [y_j]_E$ $(p \geq 2)$. Because the set $Dec_F([y]_F)$ contains at least one decision class, for each decision class $D_i \in Dec_F([y]_F)$, we have $D_i \cap [y]_F = \bigcup_{j=1}^{p} (D_i \cap [y_j]_E)$. Then, the intersection $D_i \cap [y]_F$ is either identical to an intersection $D_i \cap [y_j]_E$ by just one equivalence class $[y_j]_E$, or union of plural non-empty intersections $D_i \cap [y_j]_E$. This implies the inequality $|Dec_F([y]_F)| \leq \sum_{j=1}^{p} |Dec_E([y_j]_E)|$, which concludes $N_E \geq N_F$.

2. Suppose the given decision table is consistent and we have $R_E \subset R_F$. By this condition, for any $y \in U$ with $[y]_F = \bigcup_{j=1}^{p} [y_j]_E$ $(p \geq 2)$, $y \in D_i$ implies $y \in \underline{F}(D_i)$, and therefore we have $Dec_F([y]_F) = \{D_i\}$ by Eq. (7.19). Thus, similar to the case 1, the following inequality $1 = |Dec_F([y]_F)| < \sum_{j=1}^{p} |Dec_E([y_j]_E)|$ holds. This concludes $N_E > N_F$, and therefore we have $ACov(E) < ACov(F)$. □

Theorem 7.3 guarantees that relative reducts that provide rougher partitions receive better evaluations than those that provide finer partitions.

Table 7.4 Average of coverage of relative reducts in Table 7.1

Relative reduct	$ACov(\cdot)$
$\{c_1, c_8\}$	$\frac{3}{5}$
$\{c_4, c_8\}$	$\frac{3}{4}$
$\{c_5, c_8\}$	$\frac{3}{5}$
$\{c_2, c_3, c_8\}$	$\frac{3}{7}$

Combining Theorems 7.2 and 7.3, we can evaluate relative reducts of a given decision table by calculating the average coverage of decision rules constructed from relative reducts. Therefore, we propose to use the average coverage of decision rules constructed from relative reducts as an evaluation criterion of relative reducts based on roughness of partitions.

Example 7.5 We evaluate relative reducts of Table 7.1 by calculating the average coverage of all decision rules constructed from these relative reducts. In the case of the relative reduct $A = \{c_1, c_8\}$, as we have illustrated at the first part of this section, the partition U/R_A consists of five equivalence classes; $\{x_1\}$, $\{x_2, x_6\}$, $\{x_3, x_4\}$, $\{x_5\}$, and $\{x_7\}$. For each equivalence class, the set $Dec_A(\cdot)$ is as follows:

$$Dec_A(\{x_1\}) = \{D_1\}, \quad Dec_A(\{x_2, x_6\}) = \{D_3\}, \quad Dec_A(\{x_3, x_4\}) = \{D_2\},$$
$$Dec_A(\{x_5\}) = \{D_1\}, \quad Dec_A(\{x_7\}) = \{D_2\}.$$

Thus, the number N_A is 5 by Eq. (7.18), and therefore, the evaluation score of $A = \{c_1, c_8\}$ by the average of coverage is $ACov(A) = \frac{3}{5}$.

As we have seen in Example 7.3, the relative reduct $A = \{c_1, c_8\}$ generates actually five decision rules, and the average of coverage of these five rules is

$$\frac{1}{5}\left(\frac{1}{2} + 1 + \frac{2}{3} + \frac{1}{2} + \frac{1}{3}\right) = \frac{3}{5},$$

it is identical to the score $ACov(A)$.

Table 7.4 shows the average of coverage of each relative reduct. From this result, we conclude that the relative reduct $\{c_4, c_8\}$ is the best one that provides the most rough partition and correct approximation of decision classes.

7.4.2 Example

To demonstrate evaluation of relative reducts by the proposed method, we apply the method to Zoo dataset [8] in UCI Machine Learning Repository [26].

Zoo dataset consists of 101 samples and 17 attributes with discrete values. We use the attribute *type* as the decision attribute, and the remaining 16 attributes (*hair, feath-*

Table 7.5 Evaluation results of relative reducts in Zoo dataset

Number of attributes	Number of relative reducts	Maximum score of evaluation	Minimum score of evaluation	Average of evaluation
5	7	0.35	0.259	0.306
6	18	0.318	0.189	0.260
7	8	0.233	0.167	0.201
Total	33	0.35	0.167	0.260

ers, eggs, milk, airborne, aquatic, predator, toothed, backbone, breathes, venomous, fins, legs, tail, domestic, and *catsize*) as condition attributes. The decision attribute *type* provides seven decision classes corresponding to kinds of animals: mammal, bird, reptile, fish, amphibian, insect, and other invertebrates. Note that the decision table based on the Zoo dataset is consistent.

Consequently, there are 33 relative reducts in Zoo dataset. These include 7 relative reducts that consist of five attributes, 18 that consist of six, and 8 that consist of seven.

Table 7.5 presents the experimental results of evaluation of relative reducts consisting of five, six, and seven attributes. The maximum, minimum, and average scores of evaluation of the 33 relative reducts are 0.35, 0.167, and 0.260, respectively.

These results indicate that the smaller the number of attributes in a relative reduct, the higher the score of evaluation in general. However, even when the number of attributes is identical, big differences occur between the scores of evaluation of relative reducts. For example, the relative reduct with the highest evaluation, 0.35, is $\{milk, aquatic, backbone, fins, legs\}$, and this relative reduct generates 20 decision rules; thus, we have the score $\frac{7}{20} = 0.35$. In other words, because the decision table is consistent, this relative reduct divides 101 samples into 20 equivalence classes. In addition, even though the number of attributes is the same, the relative reduct with the lowest score of evaluation, 0.259, is $\{eggs, aquatic, toothed, legs, catsize\}$, which constructs 27 equivalence classes. Moreover, the worst relative reduct, with the evaluation score 0.167, is $\{egg, aquatic, predator, breathes, venomous, legs, catsize\}$, which constructs 42 equivalence classes.

Decision rules constructed from the best relative reduct $\{milk, aquatic, backbone, fins, legs\}$ represent characteristics of each decision class very well. For example, unlike the decision class "mammal", which is directly identified by the attribute *milk*, there is no attribute that identifies the decision class "fish" directly. However, the decision class fish, which consists of 13 kinds of fishes, is described just one decision rule:

- $[milk = no] \wedge [aquatic = yes] \wedge [backbone = yes] \wedge [fins = yes] \wedge [legs = 0]$ $\rightarrow [type = \text{fish}],\ Cer = 1.0,\ Cov = 1.0.$

Note that the attribute *fins* alone does not identify the decision class fish, because aquatic mammals in Zoo dataset such as dolphin, porpoise, and seal also have fins. Similarly, the decision class "bird", which consists of 20 kinds of birds, is described by two decision rules:

- $[milk = no] \wedge [aquatic = no] \wedge [backbone = yes] \wedge [fins = no] \wedge [legs = 2]$ $\rightarrow [type = \text{bird}]$, $Cer = 1.0$, $Cov = 0.7$,
- $[milk = no] \wedge [aquatic = yes] \wedge [backbone = yes] \wedge [fins = no] \wedge [legs = 2]$ $\rightarrow [type = \text{bird}]$, $Cer = 1.0$, $Cov = 0.3$.

These two decision rules describe the fact that all birds in Zoo dataset are characterized as non-mammal vertebrates with two legs and no fins.

7.4.3 Conclusion of Section 7.4

In this section, we reviewed a method of evaluating relative reducts of a given decision table by averages of coverage of decision rules constructed from them. The proposed method is based on the roughness and correctness of partitions based on relative reducts and evaluates relative reducts that provide roughest and most correct approximations to be better. Moreover, when we evaluate a relative reduct by the proposed method, we do not need to calculate actual scores of coverage of decision rules constructed from the relative reduct; we just need to know how many decision rules are generated from the relative reduct. Experimental results of the proposed method indicate that even when the number of attributes making up relative reducts is identical, evaluation results based on the proposed method may be quite different, and decision rules generated from a good relative reduct in the sense of the proposed evaluation represent characteristics of decision classes very well. Thus, we believe that the proposed method is very important and useful as an evaluation criterion for relative reducts.

7.5 An Example of Applications—Rough-Set-Based DNA Data Analysis

In this section, we review an approach of rough-set-based DNA data analysis. This section is based on the first author's paper [16].

7.5.1 Background

DNA microarray technology has enabled us to monitor the expression levels of thousands of genes simultaneously under certain conditions, and has been yielded various applications in the field of disease diagnosis [30], drug discovery [7], and toxicological research [27]. Among them, cancer informatics based on gene-expression data is an important domain that has promising prospects for both clinical treatment and biomedical research. One of the key issues in this domain is to discover biomarker genes for cancer diagnosis from a massive amount of gene-expression data by using a bioinformatics approach called gene selection. A typical gene-selection approach is a statistical test such as t-test and ANOVA [6]. This approach detects differentially expressed genes between groups of samples obtained from different cells/tissues. Most of the statistical tests assume that the expression values of each gene across the samples follow a prior probability distribution; hence a sufficiently large number of samples are required to obtain statistically reliable results. Rough set theory [21, 22] provides a theoretical basis for set-theoretical approximation and rule generation from categorical data. Computation of relative reducts is one of the hottest and most important research topics in rough set theory as a basis for rule generation. Relative reducts are minimal sets of attributes for correctly classifying all samples to those classes. We then expect that computation of relative reducts from gene-expression data is useful for discovering biologically-meaningful information such as biomarker candidates for cancer diagnosis. Because computing all relative reducts of the given data requires very high computational cost, there have been many proposals of heuristic algorithms to compute some of the candidates of relative reducts [5, 10, 11, 14, 29]. Kudo and Murai proposed attribute-reduction algorithms to compute as many relative reducts as possible from a decision table with numerous condition attributes [13] (reviewed in Sect. 7.3). They also proposed an evaluation criterion of relative reducts that evaluates the usefulness of relative reducts from the viewpoint of decision-rule generation [12] (reviewed in Sect. 7.4). In this section, Kudo and Murai's heuristic attribute reduction algorithms [13] and a criterion of relative reducts [12] are introduced for gene-expression data analysis. We use these algorithms and criterion in two gene-expression datasets, breast cancer [28] and leukemia [1], and discuss the extracted decision rules from these datasets and their biological meanings. The experimental results indicate that the method used in this section can identify differentially expressed genes between different classes in gene-expression datasets and that it can be useful for gene-expression data analysis.

7.5.2 Methodology

The method we use in this section to extract decision rules from gene-expression data based on rough set theory consists of the following three components: (1) Extraction of as many relative reducts as possible from gene-expression data; (2) Selection of relative reducts in accordance with an evaluation criterion of relative reducts; (3) Construction of decision rules from the selected relative reducts. In the following section, in terms of the method we use in this paper, we introduce heuristic attribute-reduction algorithms for generating as many relative reducts as possible [13] used in the first step of the above method and a criterion for evaluating the usefulness of relative reducts [12] as in the second step.

7.5.3 Datasets

To evaluate the usefulness of our method, we use two gene-expression datasets: breast cancer [28] and leukemia [1]. Both of them are two-class datasets. The leukemia dataset is composed of the gene-expression values for 12,582 genes in 24 acute lymphocytic leukemia (ALL) samples and 28 acute myeloid leukemia (AML) samples. The breast cancer dataset includes the gene-expression values for 7,129 genes in 25 positive and 24 negative samples. For each dataset, the expression values from each gene are linearly normalized to have mean 0 and variance 1. Subsequently, they are discretized into six bins $(-3, -2, -1, 1, 2, 3)$ by uniformly dividing the difference between the maximum and the minimum in the normalized data and into one bin that represents the lack of gene-expression values. Discretized positive values represent that the genes are up-regulated, while negative values represent that genes are downregulated.

7.5.4 Results and Discussion

7.5.4.1 Parameters

Our method was implemented in Java on a Linux workstation (CPU: Intel Xeon X5460 (3.16 GB) x2, Memory: 8 GB, HDD: 160 GB, OS: SUSE Linux 10.1). All experiments were conducted with the following parameters: base size $b = 10$, size limitation $L = 25$, and number of iterations $I = 100$.

7.5.4.2 Classification Accuracy

First, we evaluate the classification accuracy of our method. The evaluation is conducted by Leave-One-Out Cross Validation (LOOCV). In LOOCV, first, we extract one sample as a test sample from the dataset and generate rules using the remaining samples. Second, we check whether the test sample is correctly classified by the rules. These processes are repeated for all samples. Finally, we calculate the rate of correctly classified samples. The classification accuracy is compared to those of the two salient classifiers, decision tree (C4.5) and support vector machine (SVM). Table 7.6 shows the results of LOOCV on our method, C4.5, and SVM. For the breast cancer dataset, our method exhibits the similar classification ability with C4.5 and SVM. For the leukemia dataset, the classification ability of our method exceeds greatly that of C4.5.

Extract rules for Breast cancer dataset are as follows:

- $[CRIP1 \geq -2] \rightarrow [\text{Class} = \text{Positive}]$, $Cer = 0.76$, $Cov = 0.64$,
- $[CRIP1 = -3] \rightarrow [\text{Class} = \text{Negative}]$, $Cer = 0.95$, $Cov = 0.79$.

Table 7.6 Comparison of classification accuracy of the proposed method, decision tree, and SVM

	Classification accuracy (%)	
	Breast cancer	Leukemia
Proposed method	82.86	92.86
Decision tree (C4.5)	83.67	73.61
SVM (linear kernel)	87.75	97.22

Note that the gene *M34715_at* was removed in combining generated decision rules from the best relative reduct {*CRIP1*, *M34715_at*} because, in this case, *M34715_at* was used only for classifying one positive subject with *CRIP1* = −3.

Extracted rules for Leukemia dataset are as follows:

- $[POU2AF1 \geq -2] \rightarrow [\text{Class} = \text{ALL}]$, $Cer = 1$, $Cov = 0.88$,
- $[POU2AF1 = -3] \rightarrow [\text{Class} = \text{AML}]$, $Cer = 1$, $Cov = 1$.

7.5.4.3 Biological Meanings of Extracted Rules

Next, we discuss the biological meanings of the best results by applying our method 10 times for each dataset. In these experiments, we used the same parameter settings with the comparison experiments. The best relative reducts of two datasets are as follows: (1) Breast cancer dataset: {*CRIP1*, *M34715_at*}, $ACov = 0.08$ (=2/26). (2) Leukemia dataset: {*POU2AF1*}, $ACov = 0.29$ (=2/7), where the score $ACov$ is the average of coverage of decision rules generated from the relative reduct defined by Eq. (7.25). For example, the relative reduct {*POU2AF1*} of leukemia dataset generates 7 decision rules from 2 classes, i.e., AML and ALL; hence $ACov$ score of the relative reduct {*POU2AF1*} is 2/7 (=0.29). We extracted rules from each dataset by performing the following three steps: (1) generating all decision rules by the best relative reduct of each dataset, (2) removing decision rules that contain null values in the antecedents, and (3) combining the generated decision rules as long as possible by interpreting the meanings of decision rules. As a result, we obtained the rules for each dataset.

The extracted rules are evaluated on the basis of known biological findings. To this end, we investigate the functions of genes in the rules by reference to a genetic disease database (OMIM) [20] and a protein sequence database (Swiss-Prot) [25]. For the breast-cancer dataset, the samples can be discriminated into a true class with an accuracy of 88 percent according to the expression level of the Cystein-rich intestinal protein 1 (*CRIP1*). *CRIP1* is a transcription-factor gene that induces apoptosis in cancer cells. Interestingly, this gene has been identified as a novel biomarker of human breast cancer in recent studies [17, 18]. In the extracted rule, we can see that the *CRIP1* expression is more upregulated in the positive samples. Indeed, this is consistent with the recent findings by Ma et al. [18] that *CRIP1* in human breast cancer was overexpressed, compared to normal breast tissue in in situ experiments.

For the leukemia dataset, all samples can be perfectly discriminated by the expression level of the POU class 2 associating factor 1 (*POU2AF1*). *POU2AF1* is known as a gene responsible for leukocyte differentiation. In Swiss-Prot, we can see the description that "a chromosomal aberration involving *POU2AF* may be a cause of a form of B-cell leukemia." Namely, it suggests that this gene can be inactivated/down-regulated in lymphocytic leukemia, such as ALL. In contrast, it should be noted that *POU2AF1* in the extracted rule shows a weaker expression in AML than ALL. At present, the detailed role of *POU2AF1* in AML has not been revealed [9], whereas we expect that its biological relevance will be unveiled by experimental biologists in the near future.

7.5.5 Conclusion of Section 7.5

In this section, we reviewed a combined method of heuristic attribute reduction and evaluation of relative reducts in rough set theory for gene-expression data analysis [16]. This method is based on a heuristic attribute-reduction algorithm for generating as many relative reducts as possible and a criterion for evaluating the usefulness of relative reducts. The proposed method was applied to two geneexpression datasets: breast cancer and leukemia. In the comparison of the proposed method with C4.5 and SVM, the proposed method showed good classification accuracy that is comparable to the results of SVM and considerably exceeds that of C4.5. The experimental results also showed that the proposed method can identify differentially expressed genes among different classes in gene-expression datasets. For the breast cancer dataset, the proposed method extracted decision rules regarding a gene that has been identified as a novel biomarker of human breast cancer in recent studies. For the leukemia dataset, rules about a gene responsible for leukocyte differentiation were extracted. Thus, these results indicate a possibility that the proposed method can be a useful tool for gene-expression data analysis.

7.6 Summary

In this chapter, we reviewed an approach of rough-set-based data analysis by the authors. As we explained in Introduction, this approach is mainly based on (1) reduction of decision table [13], and (2) evaluation of relative reduct by roughness of partition [12].

Reduction of decision table and iteration of extracting as many relative reducts as possible from reduced decision tables proved a method to obtain relative reducts from a decision table with numerous condition attributes. This is based on the theoretical basis of the reduced decision table.

We think that the evaluation method of relative reduct based on roughness of partitions is a useful method for selecting better relative reducts for rough-set-based data analysis. This approach is based on computing the average of coverage of decision rules obtained from relative reducts, and the higher the score of the relative reduct, we can obtain useful decision rules for data analysis.

Actually, this approach is applied to DNA data analysis [16]. Extraction of relative reducts as many as possible from DNA data and selection of the best relative reduct based on roughness of partitions enable us to obtain a good candidate for analying DNA datasets. Combining obtained decision rules based on the good candidate, decision rules with interesting biological meanings are obtained.

References

1. Armstrong, S.A., et al.: Nat. Genet. **30**(1) (2002) [PMID:11731795]
2. Bazan, J.G., Skowron, A., Synak, P.: Dynamic reducts as a tool for extracting laws from decisions tables. In: Methodologies for Intelligent Systems. LNCS, vol. 869, pp. 346–355. Springer (1994)
3. Bazan, J.G.: Dynamic reducts and statistical inference. In: Proceedings of IPMU'96, pp. 1–5 (1996)
4. Bazan, J.G.: A comparison of dynamic and non-dynamic rough set methods for extracting laws from decision tables. In: Rough Sets in Knowledge Discovery, pp. 321–365. Physica-Verlag (1998)
5. Chouchoulas, A., Shen, A.: Rough set-aided keyword reduction for text categorization. Appl. Artif. Intell. **15**(9), 843–873 (2001)
6. Cui, X., Churchill, G.A.: Genome Biol. **4**(4) (2003) [PMID:12702200]
7. Debouck, C., Goodfellow, P.N.: Nat. Genet. **21**(1) (1999) [PMID:9915501]
8. Forsyth, R.: Zoo Data Set (1990). http://archive.ics.uci.edu/ml/datasets/Zoo. Accessed 15 May 1990
9. Gibson, S.E., et al.: Am. J. Clin. Pathol. **126**(6) (2006) [PMID:17074681]
10. Guan, J.W., Bell, D.A.: Rough computational methods for information systems. Artif. Intell. **105**, 77–103 (1998)
11. Kryszkiewicz, M., Lasek, P.: FUN: fast discovery of minimal sets of attributes functionally determining a decision attribute. Trans. Rough Sets **9**, 75–95 (2008)
12. Kudo, Y., Murai, T.: An evaluation method of relative reducts based on roughness of partitions. Int. J. Cogn. Inform. Nat. Intell. **4**(2), 50–62 (2010)
13. Kudo, Y., Murai, T.: An attribute reduction algorithm by switching exhaustive and heuristic computation of relative reducts. In: Proceedings of IEEE GrC2010, pp. 265–270. IEEE (2010)
14. Kudo, Y., Murai, T.: Heuristic algorithm for attribute reduction based on classification ability by condition attributes. J. Adv. Comput. Intell. Intell. Inform. **15**(1), 102–109 (2011)
15. Kudo, Y., Murai, T.: A note on attribute reduction from large-scale data based on rough sets. In: Proceedings of the 28th Fuzzy System Symposium, pp. 759–760 (2012) (in Japanese)
16. Kudo, Y., Okada, Y.: A heuristic method for discovering biomarker candidates based on rough set theory. Bioinformation **6**(5), 200–203 (2011)
17. Liu, S., et al.: Mol. Cancer Res. **2** (2004) [PMID:15328374]
18. Ma, X.J., et al.: Proc. Natl. Acad. Sci. USA **100**(59) (2003) [PMID:12714683]
19. Mori, M., Tanaka, H., Inoue, K. (eds.): Rough Sets and KANSEI—Knowledge Acquisition and Reasoning from KANSEI Data, Kaibundo (2004) (in Japanese)
20. OMIM. http://www.nslij-genetics.org/search_omim.html
21. Pawlak, Z.: Rough sets. Int. J. Comput. Inf. Sci. **11**, 341–356 (1982)

22. Pawlak, Z.: Rough Sets: Theoretical Aspects of Reasoning about Data. Kluwer Academic Publisher (1991)
23. Polkowski, L.: Rough Sets: Mathematical Foundations, Advances in Soft Computing. Physica-Verlag (2002)
24. Skowron, A., Rauszer, C.M.: The discernibility matrix and functions in information systems. In: Słowiński, R. (ed.) Intelligent Decision Support: Handbook of Application and Advance of the Rough Set Theory, pp. 331–362. Kluwer Academic Publishers (1992)
25. Swiss-Prot. http://au.expasy.org/sprot/
26. UCI Machine Learning Repository. http://archive.ics.uci.edu/ml/
27. Vrana, K.E., et al.: Neurotoxicology **24**(3) (2003) [PMID:12782098]
28. West, M., et al.: Proc. Natl. Acad. Sci. USA. **98**(20) (2008) [PMID:11562467]
29. Yao, Y.Y., et al.: A model of user-oriented reduct construction for machine learning. Trans. Rough Sets **8**, 332–351 (2008)
30. Yoo, S.M., et al.: J. Microbiol. Biotechnol. **19**(7) (2009) [PMID:19652509]
31. Zhang, J., Wang, J., Li, D., He, H., Sun, J.: A new heuristic reduct algorithm based on rough sets theory. In: Proceedings of WAIM2003. LNCS, vol. 2762, pp. 247–253. Springer (2003)

Chapter 8
Bilattice Tableau Calculi with Rough Set Semantics

Yotaro Nakayama, Seiki Akama, and Tetsuya Murai

Contents

Abstract A bilattice is an algebraic lattice that can represent both degrees of truth and epistemic state with the amount of information for a proposition. Rough sets are an approximation space in terms of an equivalence relation and adopted to manage uncertain and inconsistent information. In the paper, we propose a construction of a bilattice with an approximation space of rough sets. The information system of rough sets can be represented with decision logic and this can be reconstructed with a deduction system based on a bilattice. We discuss a pair of rough sets as bilattice elements and construct a deductive system with tableau calculi with a consequence relation with four-valued semantics and showed a sketch of completeness theorem.

Keywords Many-valued logic · Tableau calculi · Decision logic · Variable precision rough set · Knowledge representation

Y. Nakayama (✉)
BIPROGY Inc., 1-1-1, Toyosu, Koto-ku, Tokyo 135-8560, Japan
e-mail: yotaro.nakayama@biprogy.com

S. Akama
C-Republic, Inc., 1-20-1 Higashi-Yurigaoka, Asao-ku, Kawasaki 215-0012, Japan
e-mail: akama@jcom.home.ne.jp

T. Murai
Chitose Institute of Science and Technology, 758-65 Bibi, Chitose 066-865, Japan
e-mail: t-murai@photon.chitose.ac.jp

J. M. Abe (ed.), *Advances in Applied Logics*, Intelligent Systems Reference Library 243,
https://doi.org/10.1007/978-3-031-35759-6_8

8.1 Introduction

The notion of bilattices for Belnap's four-valued logic [1] was first introduced by Ginsberg [2] for default reasoning in artificial intelligence. Four-valued logic was originally studied by Belnap to handle both incomplete and inconsistent information expressed in a database system. Therefore, bilattices are suitable to represent inconsistent and incomplete information which is needed for an information system based on rough sets.

As for approximation space, Pawlak introduced a theory of rough sets for handling rough and coarse information [3]. Rough set theory can handle the concept of approximation by the indiscernibility relation, which is a central concept in rough set theory. The lower and the upper approximation of rough sets divide sets into three regions, namely, the positive, negative, and boundary regions. These three regions intuitively correspond to truth values of three-valued logic such as true, unknown and false. Thus, three-valued logic of rough sets has been widely studied because the third value is thought to correspond to the boundary region of rough sets [4, 5].

In this study, we propose a bilattice based on approximation space of rough sets adopting a semantic tableau as a deductive system of rough sets. We combine a bilattice as deduction basis and rough sets as elements of a bilattice. Besides, we adopt a tableau calculi for bilattices and propose a consequence relation for four-valued semantics with the order of truth-values. Bilattices are suitable to handle inconsistent and incomplete information in decision tables, as a result, they can serve as a deductive basis for decision logic. In addition, The bilattice framework enables us to represent not only a degree of truth value but also the epistemic state for information which is a topic in our research.

The deductive system of decision logic has been studied from the granular computing perspective, and in Fan et al. [6], an extension of decision logic was proposed for handling uncertain data tables by fuzzy and probabilistic methods. In Lin and Qing [7], a natural deduction system based on classical logic was proposed for decision logic in granular computing. In Avron and Konikowska [4], Gentzen-type three-valued sequent calculus was proposed for rough set theory based on non-deterministic matrices for semantic interpretation. Gentzen type axiomatization of three-valued logics based on partial semantics for decision logic is proposed in Nakayama et al. [8, 9].

As four-valued logic for rough sets, Mousavi and Jabehdar-Maralani [10] proposed a notion of relative-set for Belnap's four-valued logic and integrated a rough set as a pair of definable sets for a relative set. Vitória et al. [11] propose set-theoretical operations on four-valued sets for rough sets. The reasoning on rough sets is comprehensively studied in Akama et al. [12]. In Nakayama et al. [13], tableau calculi for four-valued logic was proposed for decision logic of rough sets.

A bilattice logics were studied by Arieli and Avron [14] and were formulated using Gentzen-type cut-free sequent calculi that correspond to certain bilattices. Muskens and Wintein studied four-valued and bilattice logics with the notion of double-barrelled consequence relation [15, 16]. Nakayama et al. [17] studied a bilattice logic for rough sets using tableau calculi with complex signs. Vitoria et al. [11] studied four-valued set approximation for rough sets in a set theoretical approach. Avron and Konikowska [4] studied three-valued logic for rough sets with the semantics using a non-deterministic matrix.

This paper is organized as follows. Section 8.2 presents an overview of rough sets and decision logic. In Sect. 8.3, we introduce Belnap's four-valued logic and bilattices and the semantic interpretation of four-valued logic along with the relationship between the regions of rough sets. In Sect. 8.4, tableau calculi *BTC* for bilattices are proposed. We adopt a signed formula for bilattice semantic tableau calculi to deal with the inference rule and validity of inference for the decision logic of rough sets. In Sect. 8.5, we showed a sketch of a soundness and completeness theorem for bilattice tableau calculi BTC. Finally, in Sect. 8.6, a summary of the study and possible directions for future work are provided.

8.2 Rough Set and Decision Logic

Rough set theory, proposed by Pawlak [3], provides a theoretical basis of sets based on approximation concepts. A rough set can be seen as an approximation of a set. It is denoted by a pair of sets, called the lower and upper approximation of the set. Rough sets are used for imprecise data handling. For the upper and lower approximations, any subset X of U can be in any of three states, according to the membership relation of objects in U. If the positive and negative regions on a rough set are considered to correspond to the truth value of a logical form, then the boundary region corresponds to ambiguity in deciding truth or falsity. Thus, it is plausible to adopt three-valued and four-valued logics for the basis for rough sets.

Rough set theory is outlined below. Let U be a non-empty finite set, called a universe of objects. If R is an equivalence relation on U, then U/R denotes the family of all equivalence classes of R, and the pair (U, R) is called a Pawlak approximation space, which is defined as follows:

Definition 8.1 Let R be an equivalence relation of the approximation space $S = (U, R)$, and X any subset of U. Then, the lower and upper approximations of X for R are defined as follows:

$$\underline{R}X = \bigcup\{Y \in U/R \mid Y \subseteq X\} = \{x \in U \mid [x]_{\mathrm{R}} \subseteq X\},$$
$$\overline{R}X = \bigcup\{Y \in U/R \mid Y \cap X \neq 0\} = \{x \in U \mid [x]_{\mathrm{R}} \cap X \neq \emptyset\}.$$

Definition 8.2 If $S = (U, R)$ and $X \subseteq U$, then the *R-positive*, *R-negative*, and *R-boundary* regions of X with respect to R are defined respectively as follows:

$$POS_R(X) = \underline{R}X$$
$$NEG_R(X) = U - \overline{R}X$$
$$BN_R(X) = \overline{R}X - \underline{R}X$$

Objects included in R-boundary is interpreted as *undefined* and *inconsistent*.

In general, targets of a decision logic are described by table-style format called information tables. Information table that was used by Pawlak [3] defined by $T = (U, A, C, D)$, where U is a finite and non-empty set of objects, A is a finite and non-empty set of attributes. C and D be subsets of a set of attribute A, $C, D \subseteq A$, and it is assumed that C is a conditional attribute and D a decision attribute.

Definition 8.3 The set of formulas of the decision logic language DL is the smallest set satisfying the following conditions:

1. (a, v), or in short a_v, is an atomic formula of *DL*, where the set of attribute constants are defined as $a \in A$ and the set of attribute value constants are $v \in V = \bigcup V_a$.
2. If φ *and* ψ are formulas of the *DL*, then $\sim\varphi$, $\varphi \wedge \psi$, $\varphi \vee \psi$, $\varphi \rightarrow \psi$, and $\varphi \equiv \psi$ are formulas.

The interpretation of *DL* is performed using the universe U in the Knowledge Representation System (*KR-system*) $K = (U, A)$ and the assignment function s, mapping from U to objects of formulas defined as follows:
$|\varphi|_S = \{x \in U : x \models_S \varphi\}$.
Formulas of *DL* are interpreted as subsets of objects consisting of a value v and an attribute a.

The semantic relations of compound formulas are recursively defined as follows:
$x \models_S a(x, v)$ iff $a(x) = v$,
$x \models_S \sim\varphi$ iff $x \not\models_S \varphi$,
$x \models_S \varphi \vee \psi$ iff $x \models_S \varphi$ *or* $x \models_S \psi$,
$x \models_S \varphi \wedge \psi$ iff $x \models_S \varphi$ *and* $S \models_S \psi$,
$x \models_S \varphi \rightarrow \psi$ iff $x \models_S \sim\varphi \vee \psi$,
$x \models_S \varphi \equiv \psi$ iff $x \models_S \varphi \rightarrow \psi$ and $s \models_S \psi \rightarrow \varphi$.

Let φ be an atomic formula of *DL*, $R \in C \cup D$ an equivalence relation, and X any subset of U, and a valuation v of propositional variables.

$$\|\varphi\|^v = \begin{cases} \mathbf{t} \text{ if } |\varphi|_S \subseteq POS_R(U/X) \\ \mathbf{f} \text{ if } |\varphi|_S \subseteq NEG_R(U/X) \end{cases} .$$

This shows that decision logic is based on bivalent logic.

8.3 Four-Valued Logic and Bilattice

8.3.1 Belnap's Four-Valued Logic

In Belnap's four-valued logic **B4**, four kinds of truth-values are used from the set **4** = {**T**, **F**, **N**, **B**}. These truth-values can be interpreted in the context of a computer, namely **T** means just told True, **F** means just told False, **N** means told neither True nor False, and **B** means told both True and False. Intuitively, **N** can be interpreted as undefined and **B** as over-defined, respectively.

Belnap outlined semantics for **B4** using the logical connectives using a notion of *set-ups* mapping atomic formulas into *FOUR* (**4**) in Fig. 8.1. A set-up can then be extended for any formula in **B4** in the following way:

$$s(A \mathbin{\&} B) = s(A) \mathbin{\&} s(B),$$
$$s(A \vee B) = s(A) \vee s(B),$$
$$s(\sim A) = \sim s(A).$$

It is also defined a concept of entailments in **B4**, such that A entails B just in case for each assignment of one of the four value to variables, the value of A does not exceed the value of **B** in **B4**, i.e., $s(A) \leq s(B)$ for each set-up s. Here, $\leq$ is defined as: **F** $\leq$ **B**, **F** $\leq$ **N**, **B** $\leq$ **T**, **N** $\leq$ **T**.

Belnap's logic was generalized by Ginsberg [2], with the notion of *bilattices*, which are algebraic structures that contain two partial orders simultaneously. In [2], a bilattice was proposed to represent default reasoning and common sense reasoning. The notion was further investigated and applied for logic programming and other purposes by Fitting [18].

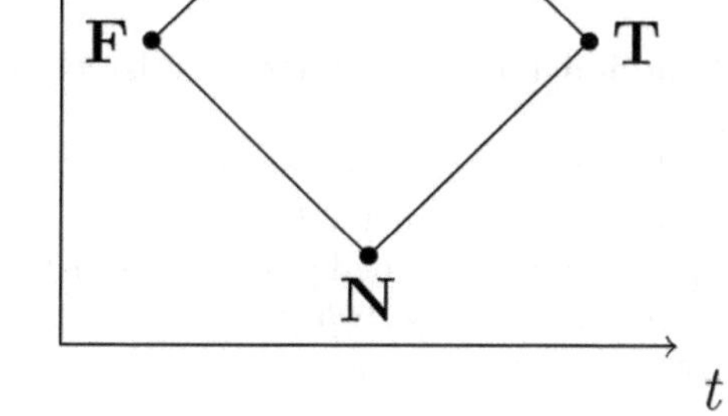

Fig. 8.1 The bilattice *FOUR*

Definition 8.4 A bilattice [2] is a structure $\mathcal{B} = (B, \leq_t, \leq_k, \neg)$ such that B is a non-empty set containing at least two elements; $(B, \leq_t)$,$(B, \leq_k)$ are complete lattices; and $\neg$ is a unary operation on B that has the following properties:
if $a \leq_t b$, then $\neg a \geq_t \neg b$.
if $a \leq_k b$, then $\neg a \leq_k \neg b$.
$\neg\neg a = a$.

Logical connective $\wedge$ and $\vee$ are interpreted with their usual meanings of "and" and "or". However $\otimes$, and $\oplus$ are understood meaning as the "consensus" and the "gullibility" or "accept all" operators, respectively. An application of $\otimes$ and $\oplus$ is provided in a logic programming language designed for distributed knowledgebases [18].

Definition 8.5 A bilattice is called distributive [2] if all the twelve possible distributive laws concerning $\wedge$, $\vee$, $\otimes$, and $\oplus$ hold. It is called interlaced [18] if each one of $\wedge$, $\vee$, $\otimes$, and $\oplus$,is monotonic with respect to both $\leq_t$ and $\leq_k$.

Definition 8.6 A structure $\mathcal{B} = (B, \leq_t, \leq_k, \neg, -)$ is a bilattice with conflation if the reduct $(B, \leq_t, \leq_k, \neg)$ is a bilattice and the conflation $- : B \rightarrow B$ is an operation satisfying:
if $a \leq_t b$, then $-a \leq_t -b$.
if $a \leq_k b$, then $-a \leq_k -b$.
$- - a = a$.

Definition 8.7 [2] Let $(L, \leq)$ be a complete lattice. The structure $L \odot L = (L \times L, \leq_t, \leq_k, \neg)$ is defined as follows:
$(y_1, y_2) \geq_t (x_1, x_2)$ iff $y_1 \geq x_1$ and $y_2 \leq x_2$
$(y_1, y_2) \geq_k (x_1, x_2)$ iff $y_1 \geq x_1$ and $y_2 \geq x_2$
$\neg(x_1, x_2) = (x_2, x_1)$

$L \odot L$ was introduced in Ginsberg [2], and later used by Fitting [19] for constructing tableau calculi for bilattices. A truth value $(x, y) \in L \odot L$ may intuitively be understood as simultaneously representing the degree of belief for an assertion, and the degree of belief against it.

In a many-valued logic, the subset of the designated value is used to define the validity of formulae and a consequence relation. Frequently, In an algebraic setting, the set of designated valued forms a filter or even a prime filter for the ordering of the truth values. For bilattices of filters and a set of designated values are the following:

Definition 8.8 (a) A bifilter of a bilattice $\mathcal{B}$ is a non-empty set $\mathcal{F} \subset \mathcal{B}$, $\mathcal{F} \neq \mathbf{B}$, such that:
$a \wedge b \in \mathcal{F}$ iff $a \in \mathcal{F}$ and $b \in \mathcal{F}$
$a \otimes b \in \mathcal{F}$ iff $a \in \mathcal{F}$ and $b \in \mathcal{F}$
(b) A bifilter $\mathcal{F}$ is called prime, if it satisfies also:
$a \wedge b \in \mathcal{F}$ iff $a \in \mathcal{F}$ or $b \in \mathcal{F}$
$a \oplus b \in \mathcal{F}$ iff $a \in \mathcal{F}$ or $b \in \mathcal{F}$

Every bifilter $\mathcal{F}$ is necessarily upward-closed w.r.t $\leq_t$ and $\leq_k$, $\{x \mid x \geq_k \mathbf{T}\}$ and $\{x \mid x \geq_t \mathbf{B}\}$ are subsets of $\mathcal{F}$. On the other hand, $\mathbf{F} \notin \mathcal{F}$, and $\mathbf{N} \notin \bot\mathcal{F}$, since $\mathcal{F} \neq$ *FOUR*.

Definition 8.9 [14] A logical bilattice is a pair $(\mathcal{B}, \mathcal{F})$, in which $\mathcal{B}$ is a bilattice, and $\mathcal{F}$ is a prime bifilter on $\mathcal{B}$.

Definition 8.10 Let $\mathcal{B}$ be a bilattice. Designated values for $\leq_k$ and $\leq_t$ of $\mathcal{B}$:
$\mathcal{D}_k(\mathcal{B}) \equiv \{x \mid x \geq_k \mathbf{T}\}$, $\mathcal{D}_t(\mathcal{B}) \equiv \{x \mid x \geq_t \mathbf{B}\}$.

8.3.2 *Rough Sets Semantics for Bilattice*

Let U be a finite non-empty set called the universe. S be an approximation space on U, and let U/R denote the set of all the equivalence classes of R. The empty set $\emptyset$ and the elements of U/R are called the elementary sets. A set which is a union of elementary sets is called the *definable set*. The family of all definable sets in approximation space *apr* is denoted by $Def(S)$. We would like to define a rough set as a pair of disjoint definable sets, i.e., given two subsets $A^+, A^- \in Def(S)$ and $A^+ \cap A^- = \emptyset$, we call the pair (A^+, A^-) a rough set in which A^+ denotes the R-positive region and A^- denotes the R-negative region of the rough set. The R-boundary region will be $U - A^+ \cup A^-$ and, if $A^+ \cup A^- = U$, the pair (A^+, A^-) will have no boundary region and will be a *crisp set*.

Relative sets are proposed to interpret rough sets as semantics to bilattices [10]. A rough set is treated as a relative set to interpret the relative truth-value concept in the form of set algebra. A pair of classical sets (A^+, A^-) is called the relative set. A^+ is the set of all objects is called the positive region of the relative set. A^- consists of all objects is called the negative region of the relative set.

According to these definitions, a relative set partitions all objects into four distinct regions:

1. If the region of all objects that belong to A^+ and do not belong to A^-. then the region corresponds to **T**.
2. If the region of all objects that belong to A^- and do not belong to A^+. then the region corresponds to **F**.
3. If the region of all objects that belong both to A^+ and A^-, then these correspond to **B**.
4. If the region of all objects that belong neither to A^+ nor A^-, then these correspond to **N**.

The relative set intersection $\cap_R$, union $\cup_R$, complement $\neg$, consensus, gullibility and conflation operations are defined as follows:

1. $(A^+, A^-) \cap_R (B^+, B^-) = (A^+ \cap B^+, A^- \cup B^-)$,
2. $(A^+, A^-) \cup_R (B^+, B^-) = (A^+ \cup B^+, A^- \cap B^-)$,
3. $\neg(A^+, A^-) = (A^-, A^+)$.

4. $(A^+, A^-) \otimes (B^+, B^-) = (A^+ \cap B^+, A^- \cap B^-)$,
5. $(A^+, A^-) \oplus (B^+, B^-) = (A^+ \cup B^+, A^- \cup B^-)$,
6. $-(A^+, A^-) = (A^+, A^-)$.

If we denote by $\mathcal{P}(U)$ the power set of U, then $(\mathcal{P}(U), \subseteq)$ will be a lattice in which meet and join operators are the classical set intersection, $\cap$, and the classical set union, $\cup$, respectively. The order of the lattice is the classical set inclusion, and the classical set complement is an order reversing involution.
Logical ordering of relative sets:
$(A^+, A^-) \subseteq_t (B^+, B^-) \Leftrightarrow A^+ \subseteq B^+, B^- \subseteq A^-$.
Approximation ordering of relative sets:
$(A^+, A^-) \subseteq_k (B^+, B^-) \Leftrightarrow A^+ \subseteq B^+, A^- \subseteq B^-$.
$\subseteq_t$ is an extension of the classical set inclusion ,which corresponds to the $\leq_t$ ordering of *FOUR*. $\subseteq_k$ is an order of information, which corresponds to the $\leq_k$ ordering of *FOUR*.

Let A^+ be the R-positive region $POS_R(A)$, A^- be the R-negative region $NEG_R(A)$ of the rough set. Now, the truth value of φ on an approximation space $S = (U, R)$ is defined as follows:

$$\|\varphi\|^v = \begin{cases} \mathbf{T} \text{ if } |\varphi|_S \subseteq POS_R(A) \\ \mathbf{F} \text{ if } |\varphi|_S \subseteq NEG_R(A) \\ \mathbf{N} \text{ if } |\varphi|_S \not\subseteq POS_R(A) \text{ and } |\varphi|_S \not\subseteq NEG_R(A) \\ \mathbf{B} \text{ if } |\varphi|_S \subseteq POS_R(A) \text{ and } |\varphi|_S \subseteq NEG_R(A) \end{cases}.$$

To handle an aspect of partiality on the decision logic, forcing relations for the partial interpretation are defined for four-valued semantic. The truth values of φ are represented by the forcing relation as follows:

$$\|\varphi\|^v = \mathbf{T} \& \text{iff } \mathcal{M} \models_v^+ \varphi \text{ and } \mathcal{M} \not\models_v^- \varphi,$$
$$\|\varphi\|^v = \mathbf{F} \& \text{iff } \mathcal{M} \not\models_v^+ \varphi \text{ and } \mathcal{M} \models_v^- \varphi,$$
$$\|\varphi\|^v = \mathbf{N} \& \text{iff } \mathcal{M} \not\models_v^+ \varphi \text{ and } \mathcal{M} \not\models_v^- \varphi,$$
$$\|\varphi\|^v = \mathbf{B} \& \text{iff } \mathcal{M} \models_v^+ \varphi \text{ and } \mathcal{M} \models_v^- \varphi.$$

A semantic relation for the model $\mathcal{M}$ is defined following Van Benthem [20], Degauquier [21] and Muskens [15]. The truth (denoted by $\models_v^+$) and the falsehood (denoted by $\models_v^-$) of the formulas of the language *DL* of the decision logic in $\mathcal{M}$ are defined inductively.

Definition 8.11 The semantic relations of $\mathcal{M} \models_v^+ \varphi$ and $\mathcal{M} \models_v^- \varphi$ are defined as follows:
$\mathcal{M} \models_v^+ \sim \varphi$ *iff* $\mathcal{M} \models_v^- \varphi$,
$\mathcal{M} \models_v^- \sim \varphi$ *iff* $\mathcal{M} \models_v^+ \varphi$,
$\mathcal{M} \models_v^+ \varphi \vee \psi$ *iff* $\mathcal{M} \models_v^+ \varphi$ *or* $\mathcal{M} \models_v^+ \psi$,
$\mathcal{M} \models_v^- \varphi \vee \psi$ *iff* $\mathcal{M} \models_v^- \varphi$ *and* $\mathcal{M} \models_v^- \psi$,
$\mathcal{M} \models_v^+ \varphi \wedge \psi$ *iff* $\mathcal{M} \models_v^+ \varphi$ *and* $\mathcal{M} \models_v^+ \psi$,

$\mathcal{M} \models_v^- \varphi \wedge \psi$ *iff* $\mathcal{M} \models_v^- \varphi$ *or* $\mathcal{M} \models_v^- \psi$,
$\mathcal{M} \models_v^+ \varphi \rightarrow \psi$ *iff* $\mathcal{M} \models_v^- \varphi$ *or* $\mathcal{M} \models_v^+ \psi$,
$\mathcal{M} \models_v^- \varphi \rightarrow \psi$ *iff* $\mathcal{M} \models_v^+ \varphi$ *and* $\mathcal{M} \models_v^- \psi$.

The symbol $\sim$ denotes strong negation, which is interpreted as true if the proposition is false. Since validity in **B4** is defined in terms of truth preservation, the set of designated values is $\{\mathbf{T}, \mathbf{B}\}$ of **4**.

8.4 Bilattice-Based Tableau Calculi

Semantic tableaux can be regarded as a variant of Gentzen systems; see Smullyan [22]. The tableau calculus is used as the proof method for both classical and non-classical logics in Akama [23] and Priest [24]. The main advantage of the use of the tableau calculus is that proofs in tableau calculi are easy to understand. In addition, it is possible to provide a comprehensive argument of completeness proof.

To accommodate the Gentzen system to bilattice logics, we need some concepts of partial semantics. In the Beth tableau, It is assumed that V is a partial valuation function assigning to a formula P the values **F** or **T**. We can then set $V(P) = \mathbf{T}$ for P on the left-hand side and $V(P) = \mathbf{F}$ on the right-hand side in an open branch of tableaux.

First, we obtain the following concept of consequence relation (8.1) for classical logic. Let *Pre* is $p_1, p_2 \ldots, p_m$ and *Cons* is $q_1, q_2 \ldots q_n$, then the consequence relation is said to be valid:

$$\text{For all } V, \text{ if } V(Pre) = \mathbf{T} \text{ then } V(Cons) = \mathbf{T} \tag{8.1}$$

We use the notion of signed formula. If φ is a formula, then $T\varphi$ and $F\varphi$ are signed formulas. $T\varphi$ reads φ is *provable* and $F\varphi$ reads φ is *not provable*, respectively. If S is a set of signed formulas and α is a signed formula, then we simply define $\{S, \alpha\}$ for $S \cup \{\alpha\}$. As usual, a tableau calculus consists of axioms and reduction rules. Let p be an atomic formula and φ and ψ be formulas.

Fitting [19] proposed an extension of semantic tableau by Smullyan. Let a bilattice $\mathcal{B}$ be distributive and it will be of the form $L_1 \odot L_2$ for lattices L_1 and L_2. It is convenient to make use of this representation. Remember, in Smullyan's system one works backwards. To establish that X is valid one begins by assuming it is not, FX, and one tries for a contradiction. Likewise now, if X is given values in the bilattice $L_1 \odot L_2$, and we want to establish the 'evidence for' X is at least $\mathbf{T}$, we begin by assuming it isn't, and try for a contradiction. Thus, we introduce a family of signs F for each $\mathbf{T} \in L_1$, and we think of FX as asserting that the valuation of X in $L_1 \odot L_2$ has a first component that is $\not\geq \mathbf{T}$. Then a contradiction based on FX will establish that the evidence for X is at least a. Similarly TX will mean the valuation of X has a second component that is $\not\geq \mathbf{F}$, where $\mathbf{F} \in L_2$.

Table 8.1 Formulas of type α and β

α	α_1	α_2	β	β_1	β_2
$T(\varphi \wedge \psi)$	$T\varphi$	$T\psi$	$T(\varphi \vee \psi)$	$T\varphi$	$T\varphi$
$F(\varphi \vee \psi)$	$F\varphi$	$F\varphi$	$F(\varphi \wedge \psi)$	$F\varphi$	$F\varphi$
$T(\varphi \oplus \psi)$	$T\varphi$	$T\varphi$	$T(\varphi \otimes \psi)$	$T\varphi$	$T\varphi$
$F(\varphi \oplus \psi)$	$F\varphi$	$F\varphi$	$F(\varphi \otimes \psi)$	$F\varphi$	$F\varphi$
$T(\sim \varphi)$	$F\varphi$	$F\varphi$			
$F(\sim \varphi)$	$T\varphi$	$T\varphi$			
$T(-\varphi)$	$T\varphi$	$T\varphi$			
$F(-\varphi)$	$F\varphi$	$F\varphi$			

Definition 8.12 For a valuation v in $\mathcal{B}$, we extend its action to map signed formulas to classical truth values, true or false, as follows:

1. $\mathcal{B} \models^+ \varphi$ is true if $v(\varphi) = \langle x_1, x_2 \rangle$ and $x_2 \not\geq \mathbf{F}$;
2. $\mathcal{B} \models^- \varphi$ is true if $v(\varphi) = \langle x_1, x_2 \rangle$ and $x_1 \not\geq \mathbf{T}$;

We classify signed formulas into *conjunctive* or α formulas, with components α_1 and α_2, and into *disjunctive* or β formulas, with components β_1 and β_2. These are given in Table 8.1.

Fitting [18] studied the connective called *conflation* corresponding to a top-bottom symmetry on Bilattice. The following equations (DeMorgan laws for conflation) hold in any bilattice:

$-(x \wedge y) = -x\wedge,\ -y - (x \vee y) = -x \vee -y,\ -(x \otimes y) = -x \oplus -y,\ -(x \oplus y) = -x \otimes -y.$

In addition, if the bilattice is bounded, then $-\mathbf{T} = \mathbf{T}$, $-\mathbf{F} = \mathbf{T}$, $-\mathbf{B} = \mathbf{N}$, $-\mathbf{N} = \mathbf{B}$.

The tableau rules for a propositional classical logic TCL are:

Axiom:

(ID) $\{Tp, Fp\}$

Tableau rule:

$$\frac{S, T(\sim\varphi)}{S, F\varphi}\ (T{\sim}) \qquad \frac{S, F(\sim\varphi)}{S, T\varphi}\ (F{\sim})$$

$$\frac{S, T(\varphi \wedge \psi)}{S, T\varphi, T\psi}\ (T\wedge) \qquad \frac{S, F(\varphi \wedge \psi)}{S, F\varphi;\ S, F\psi}\ (F\wedge)$$

$$\frac{S, T(\varphi \vee \psi)}{S, T\varphi;\ S, T\psi}\ (T\vee) \qquad \frac{S, F(\varphi \vee \psi)}{S, F\varphi, F\psi}\ (F\vee)$$

$$\frac{S, T(\varphi \to \psi)}{S, F\varphi;\ S, T\psi}\ (T{\to}) \qquad \frac{S, F(\varphi \to \psi)}{S, T\varphi, F\psi}\ (F{\to})$$

A proof of a formula φ is shown with a closed tableau for $F\varphi$. A tableau is a tree constructed by the above reduction rules. A tableau is closed if each branch is closed.

A branch is closed if it contains the axioms of the form (ID) in classical logic. We write $\vdash_{BCT} \varphi$ to mean that φ is provable in TCL.

Theorem 8.1 *The logic for the consequence relation (8.1) is axiomatized by the Bilattice-based classical tableau calculus BCT.*

The tableau calculus BCT* is extended BCT excluded rules of $(T\sim)$ and $(F\sim)$ by introducing axioms of the principle of explosion EFQ (ex falso quodlibet) and the excluded middle EM, and following rules:

(EFQ)	$\{Tp, T{\sim}p\}$
(EM)	$\{Fp, F{\sim}p\}$

$$\frac{S, T(\varphi \otimes \psi)}{S, T\varphi, T\psi}\ (T\otimes) \qquad \frac{S, F(\varphi \otimes \psi)}{S, F\varphi;\ S, F\psi}\ (F\otimes)$$

$$\frac{S, T({\sim}(\varphi \otimes \psi))}{S, T({\sim}\varphi), T({\sim}\psi)}\ (T{\sim}\otimes) \qquad \frac{S, F({\sim}(\varphi \otimes \psi))}{S, F({\sim}\varphi);\ S, F({\sim}\psi)}\ (F{\sim}\otimes)$$

$$\frac{S, T(\varphi \oplus \psi)}{S, T\varphi;\ S, T\psi}\ (T\oplus) \qquad \frac{S, F(\varphi \oplus \psi)}{S, F\varphi, F\psi}\ (F\oplus)$$

$$\frac{S, T({\sim}(\varphi \oplus \psi))}{S, T({\sim}\varphi);\ S, T({\sim}\psi)}\ (T{\sim}\oplus) \qquad \frac{S, F({\sim}(\varphi \oplus \psi))}{S, F({\sim}\varphi), F({\sim}\psi)}\ (F{\sim}\oplus)$$

$$\frac{S, T({\sim}(\varphi \wedge \psi))}{S, T({\sim}\varphi);\ S, T({\sim}\psi)}\ (T{\sim}\wedge) \qquad \frac{S, F({\sim}(\varphi \wedge \psi))}{S, F({\sim}\varphi), F({\sim}\psi)}\ (F{\sim}\wedge)$$

$$\frac{S, T({\sim}(\varphi \vee \psi))}{S, T({\sim}\varphi), T({\sim}\psi)}\ (T{\sim}\vee) \qquad \frac{S, F({\sim}(\varphi \vee \psi))}{S, F({\sim}\varphi);\ S, F({\sim}\psi)}\ (F{\sim}\vee)$$

$$\frac{S, T({\sim}(\varphi \rightarrow \psi))}{S, T\varphi, T({\sim}\psi)}\ (T{\sim}\rightarrow) \qquad \frac{S, F({\sim}(\varphi \rightarrow \psi))}{S, F\varphi;\ S, F({\sim}\psi)}\ (F{\sim}\rightarrow)$$

$$\frac{S, T({\sim}{\sim}\varphi)}{S, T\varphi}\ (T{\sim}{\sim}) \qquad \frac{S, F({\sim}{\sim}\varphi)}{S, F\varphi}\ (F{\sim}{\sim})$$

The positive rules for $\wedge$ and $\otimes$ are identical. Both behave as classical conjunction. The difference is with respect to the negations of $p \wedge q$ and $p \otimes q$. Unlike the conjunction of classical logic, the negation of $p \otimes q$ is equivalent to $\neg p \otimes \neg q$. This follows from the fact that $p \leq_k q$ iff $\neg p \leq_k \neg q$. The difference between $\wedge$ and $otimes$ is similar.

Axioms for conflation are as follows [25]:

(C1)	$--A \leftrightarrow A$	(C6)	$A \rightarrow -\mathbf{N}$
(C2)	$-\sim A \leftrightarrow \sim - A$	(C7)	$-(A \wedge B) \leftrightarrow -A \wedge -B$
(C3)	$-\mathbf{F} \rightarrow A$	(C8)	$-(A \vee B) \leftrightarrow -A \vee -B$
(C4)	$A \rightarrow -\mathbf{T}$	(C9)	$-(A \otimes B) \leftrightarrow -A \oplus -B$
(C5)	$-\mathbf{B} \rightarrow A$	(C10)	$-(A \oplus B) \leftrightarrow -A \otimes -B$

Axioms for conflation can be transformed into rules by adopting a concept of a complex sing to formulas instead of a single sign [15–17, 26].

Next, we extend consequence relation (8.1) for Bilattice tableau calculus BTC as follows:

$$\textit{For all } V,\ \textit{if } V(Pre) \geq_k \mathbf{T} \textit{ then } V(Cons)\ \geq_k \mathbf{T}. \tag{8.2}$$

Pre and *Cons* are evaluated as at least true, respectively. Equation (8.1) and (8.2) are not different from the formulation of classical validity. However, they should be distinguished in Bilattice interpretation. The tableau calculus BTC for (8.2) is defined from BCT* without ($T{\sim}$), ($F{\sim}$), (EFQ) and (EM) as follows:

$$\begin{aligned}\mathbf{BTC} := \{&(ID),\ (T\wedge),\ (F\wedge),\ (T\vee),\ (F\vee),\ (T\rightarrow),\ (F\rightarrow),\ (T{\sim}\wedge),\ (F{\sim}\wedge),\\ &(T{\sim}\vee),\ (F{\sim}\vee),\ (T{\sim}\rightarrow),\ (F{\sim}\rightarrow),\ (T\otimes),\ (F\otimes),\ (T\oplus),\ (F\oplus),\\ &(T\ {\sim}\otimes),\ (F\ {\sim}\otimes),\ (T\ {\sim}\oplus),\ (F\ {\sim}\oplus),\ (T\rightarrow),\ (T{\sim\sim}),\ (F{\sim\sim})\}.\end{aligned}$$

(8.2) is regarded as a four-valued logic since it allows for incomplete and inconsistent valuation. As the semantics for BTC, we define the bilattice valuation function $v(p)$ for an atomic formula p as follows:

Next, we turn to specific examples. First, the simplest one. Take for both L_1 and L_2 the two-element lattice {0, 1} where $0 \leq 1$. Think of 0 as no evidence, and 1 as complete certainty. Then $L_1 \odot L_2$ is a four-element bilattice in which $\mathbf{N}$ is $\langle 0, 0\rangle$, indicating we have no evidence either for or against, and $\mathbf{B}$ is $\langle 1, 1\rangle$, indicating we are in the inconsistent situation of having full evidence both for and against. Likewise $\mathbf{F}$ is $\langle 0, 1\rangle$ and $\mathbf{T}$ is $\langle 1, 0\rangle$.

In this valuation, the law of contradiction fails since if we have $p \wedge \neg p$ in premises, both p and $\neg p$ are evaluated as $\mathbf{B}$. Additionally, the law of excluded middle fails since if we have $p \vee \neg p$ in conclusions, both p and $\neg p$ are evaluated as $\mathbf{N}$.

Now, we try to extend BTC with weak negation and weak implication. Weak implication regains the deduction theorem that some many-valued logics lack.

Here, we introduce weak negation "¬". The semantic relation for weak negation is as follows:

$$\mathcal{M} \models^+_v \neg\varphi \&\text{iff}\ \ \mathcal{M} \not\models^+_v \varphi,$$
$$\mathcal{M} \models^-_v \neg\varphi \&\text{iff}\ \ \mathcal{M} \models^+_v \varphi.$$

A semantic interpretation of weak negation is denoted as follows:

$$\|\neg\varphi\|^v = \begin{cases}\mathbf{F} \text{ if } \|\varphi\|^v = \mathbf{T}\ \textit{or}\ \mathbf{B}\\ \mathbf{T} \text{ if } \|\varphi\|^v = \mathbf{F}\ \textit{or}\ \mathbf{N}\end{cases}.$$

Weak negation can represent the absence of truth and the reading for $\neg\varphi$ is as "$\neg\varphi$ is not *true*". However, "∼" can serve as strong negation to express the verification of falsity.

Next, weak implication is defined as follows:

$$\varphi \rightarrow_w \psi =_{def} \neg\varphi \vee \psi.$$

A semantic interpretation of weak implication is defined as follows:

$$\|\varphi \rightarrow_w \psi\|^v = \begin{cases} \|\psi\|^v \text{ } if \text{ } \|\varphi\|^v \geq_t \mathbf{T} \\ \mathbf{T} \text{ } if \text{ } \|\varphi\|^v \not\geq_t \mathbf{T} \end{cases}.$$

Unlike "$\rightarrow$", weak implication satisfies the deduction theorem. This means that it can be regarded a logical implication. We can also interpret weak negation in terms of classical negation and weak implication:

$$\neg\varphi =_{def} \varphi \rightarrow_w \sim\varphi.$$

We extend BTC with weak negation and weak implication as BTC$^+$. So, we define the tableau rules for ($\neg$) and ($\rightarrow_w$) as follows:

$$\frac{S, T(\neg\varphi)}{S, F\varphi} \ (T\neg) \qquad \frac{S, F(\neg\varphi)}{S, T\varphi} \ (F\neg)$$

$$\frac{S, T(\varphi \rightarrow_w \psi)}{S, F\varphi;\ S, T\psi} \ (T{\rightarrow_w}) \qquad \frac{S, F(\varphi \rightarrow_w \psi)}{S, T\varphi, F\psi} \ (F{\rightarrow_w})$$

BTC$^+$ is interpreted as an extended four-valued logic with weak negation and weak implication.

8.5 Soundness and Completeness

In this section, the soundness and completeness theorem is shown for the tableau system BTC. A *proof* of a formula φ is a closed tableau for $F\varphi$. A tableau is a tree constructed by the reduction rules defined in the previous section. A tableau is closed if each branch is closed, where it contains the axiom of the form (ID).

We write $\vdash_{BTC} \varphi$ to mean that φ is provable in BTC. We see that φ is true iff $v(\varphi) = 1$. φ is valid, written $\models_{BTC} \varphi$, iff it is true in all four-valued models of Belnap's logic **B4**. We prove the completeness of the tableau calculus BTC with respect to Belnap's four-valued semantics. The proof strategy is similar to the way sketched in Akama [23].

Let $S = \{T\varphi_1, \cdots, T\varphi_n, F\psi_1, \cdots, F\psi_m\}$ be a set of signed formula, $\mathcal{M}$ be a four-valued with weak negation model. We say that valuation v refutes S if

$$\begin{aligned} v(\varphi_i) &= \mathbf{T}\text{\&if } T\varphi_i \in S \text{ and } F\varphi_i \notin S, \\ v(\varphi_i) &= \mathbf{F}\text{\&if } F\varphi_i \notin S \text{ and } F\varphi_i \in S, \\ v(\varphi_i) &= \mathbf{N}\text{\&if } T\varphi_i \notin S \text{ and } F\psi_i \notin S, \\ v(\varphi_i) &= \mathbf{B}\text{\&if } F\varphi_i \in S \text{ and } F\varphi_i \in S. \end{aligned}$$

A set S is *refutable* if something refutes it. If S is not refutable, it is *valid*. In addition, a set S obeying the above conditions will be called saturated. Thus for any bilattice valuation, the set of all sentences at least true under the valuation is saturated.

Theorem 8.2 (Soundness of BTC). *If* $\Gamma \models_{BTC} \varphi$ *then* $\Gamma \vdash_{BTC} \varphi$.

Proof For any formula φ in BTC, the following holds:

$\vdash_{BTC} \varphi$ iff $\models_{BTC} \varphi$

If φ is of the form of axioms, it is easy to see that it is valid. For reduction rules, it suffices to check that they preserve validity. We only show the cases of $(T{\sim}\vee)$ and $(F{\sim}\rightarrow)$.

$(T{\sim}\vee)$: We have to show that if $S, T(\sim(\varphi \vee \psi))$ is refutable then $S, T(\sim\varphi)$, $T(\sim\psi)$ is also refutable. By the assumption, there is a semantic relation Definition 8.11, in which valuation v refutes S and $\models_v^+ \sim(\varphi \vee \psi)$. This implies:

$$\begin{aligned} v(\varphi \vee \psi) \not\geq_k \mathbf{T} &\text{ iff } v(\varphi) \not\geq \mathbf{T} \text{ and } v(\psi) \not\geq \mathbf{T} \\ &\text{ iff } v(\varphi) \leq \mathbf{F} \text{ and } v(\psi) \leq \mathbf{F} \\ &\text{ iff } v(\sim\varphi) \geq \mathbf{T} \text{ or } v(\sim\psi) \geq \mathbf{T}. \end{aligned}$$

Therefore, $S, T(\sim\varphi), T(\sim\psi)$ is shown to be refutable.

$(F{\sim}\rightarrow)$: By the assumption, there is a semantic relation of the weak negation, which refutes S and $\models_v^- \sim(\varphi \rightarrow \psi)$. This implies:

$$v(\varphi \rightarrow \psi) \geq \mathbf{T} \text{ iff } v(\varphi) \leq \mathbf{F} \text{ and } v(\sim\psi) \leq \mathbf{F}.$$

Therefore, $S, F\varphi$ and $S, F(\sim\psi)$ are refutable. We can show other cases. □

We are now in a position to prove the completeness of BTC. The proof below is similar to the Henkin proof described in Akama [23], which is extended for paraconsistent logic.

A finite set of signed formulas Γ is *non-trivial* if no tableau for it is closed. An infinite set of signed formulas is non-trivial if every finite subset is non-trivial. If a set of formulas is not non-trivial, it is trivial. Every formula is provable from a trivial set.

Lemma 8.1 *A non-trivial set of signed formulas Γ_0 can be extended to a maximally non-trivial set of signed formulas Γ.*

Proof By Lindenbaum's Lemma, we can obtain a non-trivial set of signed formulas as a maximal consistent superset. See [13, 23]. □

A branch e of a tableau is called complete if for every α which occurs in a branch θ, both α_1 and α_2 occur in a tableau $\mathcal{T}$, and for every β which occurs in $theta$, at least one of β_1, β_2 occurs in θ. We call a tableau $\mathcal{T}$ completed if every branch of $\mathcal{T}$ is either closed or complete.

Theorem 8.3 *Any complete open branch of any tableau is (simultaneously) satisfiable.*

Proof Let θ be a complete open branch of a tableau $\mathcal{T}$; let S be the set of terms of θ. Then the set S satisfies the following three conditions (for every (α, β)):

H_0: No signed variable and its conjugate are both in S).

H_1: If $\alpha \in S$, then $\alpha_1 \in S$ and $\alpha_2 \in S$.

H_2: If $\beta \in S$, then $\beta_1 \in S$ or $\beta_2 \in S$. □

Lemma 8.2 *Hintikka's Lemma. Every downward saturated set S (whether finite or infinite) is satisfiable.*

Hintikka's lemma is equivalent to the statement that every Hintikka set can be extended to a saturated set.

Theorem 8.4 *(Completeness Theorem).* $\Gamma \vdash_{BTC} \psi \iff \Gamma \models_{BTC} \psi$

Proof (1) Soundness (Left to right), if $\Gamma \vdash_{BTC} \psi$ then $\Gamma \models_{BTC} \psi$. By applying Theorem 8.2, we obtain the soundness.

(2) Completeness theorem (Right to left), a contraposition is adopted. Let assume that $\Gamma \nvdash_{BTC} \psi$. Construction of a saturated set such taht (i) Γ is maximal non-trivial set and (ii) If $\Gamma \vdash \psi$ then $\psi \in \Gamma$, where Lemma 8.1 and Theorem 8.3 can be applied. Hence, $\Gamma \not\models_{BTC} \psi$. Therefore, we derive the completeness theorem. □

Consider a completed open tableau for the inference, and choose an open branch. The interpretation that the branch induces makes all the members of Γ true, and ψ false, by the completeness theorem. Therfore a completed open tableau is saturated and maximal non-trivial.

8.6 Conclusion

In this paper, we have formalized bilattice tableau calculi for rough sets. As semantics for inconsistent information tables of rough sets, we introduce bilattice to interpret with truth order and information order. Four-valued semantics with bilattices is applied to extend decision logic for interpretation of inconsistent data table with two points of view, e.g. degree of truth and amount of information. Furthermore, we extended the four-valued tableau calculi with weak negation to repair the deduction of a bilattice.

There are some topics that can be further developed. First, it is very interesting in applying another kind of proof system or deduction method to decision logic. We are also interested to apply another kind of semantics and models to interpret decision logic that includes inconsistent information. Second, we need to extend the present work for the predicate logic for decision logic, and also need to show strong completeness. Third, we need to investigate the applications of decision logic with bilattice semantics for inconsistent and incomplete information to, e.g. an agent knowledge management or a software specification.

References

1. Belnap, N.D.: A useful four-valued logic, vol. 2. Reidel Publishing (1977)
2. Ginsberg, M.L.: Multivalued logics: a uniform approach to inference in artificial intelligence. Comput. Intell. **4**, 256–316 (1988)

3. Pawlak, Z.: Rough Sets: Theoretical Aspects of Reasoning about Data. Kluwer, Dordrecht (1991)
4. Avron, A., Konikowska, B.: Rough sets and 3-valued logics. Studia Logica **90**, 69–92 (2008)
5. Ciucci, D., Dubois, D.: Three-valued logics, uncertainty management and rough sets. In: Transactions on Rough Sets XVII, Lecture Notes in Computer Science book series (LNCS, volume 8375), pp. 1–32 (2001)
6. Fan, T.F., Hu, W.C., Liau, C.J.: Decision logics for knowledge representation in data mining. In: 25th Annual International Computer Software and Applications Conference. COMPSAC, pp. 626–631 (2001)
7. Lin, Y., Qing, L.: A logical method of formalization for granular computing. In: IEEE International Conference on Granular Computing (GRC 2007), pp. 22–27 (2007)
8. Nakayama, Y., Akama, S., Murai, T.: Deduction system for decision logic based on partial semantics. In: The 11th International Conference on Advances in Semantic Processing SEMAPRO 2017 (2017)
9. Nakayama, Y., Akama, S., Murai, T.: Deduction system for decision logic based on many-valued logics. Int. J. Adv. Intell. Syst. **11**(1&2), 115–126 (2018)
10. Mousavi, A., Jabedar-Maralani, P.: Relative sets and rough sets. Int. J. Appl. Math. Comput. Sci. **11**(3), 637–653 (2011)
11. Vitoria, A., Andrzej, A.S., Maluszynski, J.: Four-valued extension of rough sets. In: International Conference on Rough Sets and Knowledge Technology RSKT, pp. 106–114 (2008)
12. Akama, S., Murai, T., Kudo, Y.: Reasoning with Rough Sets. Logical Approaches to Granularity-Based Framework. Springer, Heidelberg (2018)
13. Nakayama, Y., Akama, S., Murai, T.: Four-valued tableau calculi for decision logic of rough set. In: Knowledge-Based and Intelligent Information & Engineering Systems: Proceedings of the 22nd International Conference, KES-2018, pp. 383–392 (2018). https://doi.org/10.1016/j.procs.2018.07.272
14. Arieli, O., Avron, A.: Reasoning with logical bilattices. J. Log., Lang. Inf. **5**, 25–63 (1996)
15. Muskens, R.: On partial and paraconsistent logics. Notre Dame J. Formal Log. **40**, 352–374 (1999)
16. Wintein, S., Muskens, R.: A calculus for Belnap's logic in which each proof consists of two trees. Logique et Analyse **55**, 643–656 (2012)
17. Nakayama, Y., Akama, S., Murai, T.: Bilattice logic for rough sets. J. Adv. Comput. Intell. Intell. Inf. **24**(6), 774–784 (2020). https://doi.org/10.20965/jaciii.2020.p0774
18. Fitting, M.: Bilattices and the semantics of logic programming. J. Log. Program. **11**, 91–116 (1991)
19. Fitting, M.: Bilattices in logic programming. Proceed. Int. Symp. Multiple-Valued Log. **1**, 238–246 (1990)
20. Van Benthem, J.: Partiality and nonmonotonicity in classical logic. Logique et Analyse **29**, 225–247 (1986)
21. Degauquier, V.: Partial and paraconsistent three-valued logics. Log. Log. Philos. **25**, 143–171 (2016)
22. Smullyan, R.: First-Order Logic. Dover Books (1995)
23. Akama, S.: Nelson's paraconsistent logics. Log. Log. Philos. **7**, 101–115 (1999)
24. Priest, G.: An Introduction to Non-Classical Logic From If to Is 2nd Edition (2008)
25. Greco, G., Liang, F., Palmigiano, A., Rivieccio, U.: Bilattice logic properly displayed. Fuzzy. Sets. Syst. **363**, 138–155 (2019). https://doi.org/10.1016/j.fss.2018.05.007. Theme: Many-valued and Fuzzy Logics
26. Hahnle, R.: Automated Deduction in Multiple-Valued Logics. Oxford University Press, Oxford (1994)

Chapter 9
Optimizing the Data Loss Prevention Level Using Logic Paraconsistent Annotated Evidential Eτ

Liliam Sayuri Sakamoto, Jair Minoro Abe, Jonatas Santos de Souza, and Luiz Antonio de Lima

Contents

Abstract Currently, corporations worldwide have a problem that grows exponentially: orchestrating the organization and understanding structured and unstructured data. These unstructured data can be grouped in repositories, for isolated and random data entry, however on the other hand the data loss analysis, that is, data loss prevention, where some criteria of artificial intelligence create templates that are monitored, and that, because they are very restricted, also present contradictions and flaws. The focus of this study is to optimize this analysis by minimizing the level of data loss using the Logic Paraconsistent Annotated Evidential Eτ. With a bibliographic review on DLP—Data Loss Prevention, Logic Paraconsistent Annotated Evidential

L. S. Sakamoto (✉) · J. S. de Souza · L. A. de Lima
Paulista University, 1212 Dr. Bacelar Street, São Paulo, Brazil
e-mail: liliam.sakamoto@aluno.unip.br

J. M. Abe
Graduate Program in Production Engineering, Paulista University, São Paulo, Brazil

J. M. Abe (ed.), *Advances in Applied Logics*, Intelligent Systems Reference Library 243,
https://doi.org/10.1007/978-3-031-35759-6_9

Eτ, Artificial Intelligence techniques, and data protection. With the use of a Python program, and applied research will be carried out with data from a financial company, which presents 40% of data loss in its analysis with this process of Artificial Intelligence, in comparison with the minimization of this data loss with the use of Logic Paraconsistent Annotated Evidential Eτ to 25%, that is, a difference of 15%.

Keywords DLP—Data Loss Prevention · Logic paraconsistent Annotated evidential Eτ · Artificial intelligence

9.1 Introduction

9.1.1 General Context

It is interesting how with the evolution of operational and production processes, there was also an evolution in capturing data and transforming them into feasible information. Today, digitally structured data can be easily identified in a company, as they are within the essential systems and in organized repositories. However, how to identify all other unstructured data, both physical, isolated, and merged in social networks, text messages, emails, cloud drives, saved on flash drives or external hard drives, stored on the end users' hard drives in their endpoints, and the data that is transiting from one point to the other? [11].

The differential of each company to identify this data is in its risk appetite because the more the data and the transformed information become an asset of increasing value, the greater the importance of its security and organization. Understanding the flow of these input, transformation, output, and transit data brings security that there is active monitoring and that data loss can be nalysed through Artificial Intelligence algorithms.

Despite this type of tool being advanced in contrast to the manual analysis of data, some data may present doubts, causing problems in identifying the monitored data template. For this reason, the implementation of another analysis layer with the Logic Paraconsistent Annotated Evidential Eτ adds greater assertiveness in decision making.

A sample of a financial company that asks for confidentiality was collected, noting that the data were anonymized in compliance with the LGPD—General Data Protection Law of Brazil to test this implementation [22], only referring to a period of 1 month of data.

9.1.2 General Data Protection Law of Brazil

The protection of personal data [21], public or private, sensitive or not, is directly related to the protection of the intimacy and private life of individuals. It is worth remembering that the right to privacy is, as a rule, born with a negative aspect, that is, the right not to be molested.

According to the LGPD [22], it is the citizen's right to know, correct, and determine to whom, how, when, and how their data will be disclosed. It can be seen that the revolution brought about by the advent and diffusion of the Internet has given a new meaning to the right to privacy.

Therefore, the citizen has the right to be alone and the possibility of demanding concrete benefits, that is, demanding information, correcting it, and controlling its use. It is interesting to note that the need for legal protection of the citizen originates in the realization that the personal data that circulate on the web have an economic content, that is, that there is the possibility of commercialization of such data.

The impact of this Law, in line with technological evolution and the need for business innovation brought about by this new scenario, led directly to creating economic, social, and political effects in the country. As a result, there is a need for professionals with specialized knowledge in specific topics aimed at protecting personal data, which is called in the GDPR of the DPO (Data Protection Officer) and in the LGPD of the person in charge of personal data.

DPO—Data Protection Officer will have the challenge of coordinating and bringing companies into compliance, not only with the LGPD but with other laws that deal with the matter, such as the Consumer Protection Code, Marco Civil da Internet Law [22], Federal Constitution [2]; in addition to the laws, there are also information security standards (normative standards, among others) [22] that should guide the security practices applied. As if that were not enough, they must still know how to talk about business because aiming at all this compliance, it is necessary to map processes and understand the flow of information, people involved, and technologies to support the entire process.

A new law was recently enacted in Brazil, arousing the population's collective interest, governmental and non-governmental bodies, and business segments. The General Law for the Protection of Personal Data, Law no. 13,709 [23] of August 14, 2018, grants the Brazilian population rights and guarantees on how organizations will adapt to the collection and processing of personal data, whether by physical or digital means.

Companies are now concerned with protecting people's data due to the approval of the Brazilian Law that was named Marco Civil da Internet [23].

Brazil leads in America in terms of uniqueness in terms of awareness, regulation, new businesses, as it brings together internationally trained professionals and officially recognized on July 29, 2021, as the occupation of DPO/Data Processing Officer was included in the Brazilian Classification of Occupations (CBO) which will begin in 2022.

The Data Officer has a vital role in this mission provided for in Art. 41, and in order not to run the risk of suffering from the high fines that the LGPD will charge anyone who is not in compliance with its rules, the ideal is to find a professional to help the company in the transition.

The relevance of Data Privacy has greatly increased in recent years. The theme gained more evidence after the scandals involving social media companies accused of selling personal data to other marketing companies, there was also the case with the email of former secretary of defense Hilary Clinton, among others. This triggered serious postures in several countries in relation to information security and the result was the creation of strict laws on the subject. In Brazil, DPO professionals unite through the ANPPD—National Association of Data Privacy Professionals in Brazil.

A great example is the GDPR, the European Regulation on Privacy of Personal Data that served as a model for the Brazilian law LGPD—General Personal Data Privacy Law authorized in 2018, sanctioned by the then President of the Republic Michel Temer (PMDB) on December 14. August 2018, with full force as of 2020.

Both laws agree that, in order to adapt to companies, it is necessary to involve areas such as IT/Information Security and Legislation; for this, they bring as a requirement a professional responsible for the privacy of personal data, the Data Privacy Officer in Brazil, known worldwide as DPO—Data Protection Officer.

During this scenario, several information security professionals and lawyers began to observe this new field of activity; however, unlike lawyers, IT professionals did not have an organization that represented the class in the decisions that were being processed in the National Congress regulating the details about the role of the DPO.

Even in 2019, before the emergence of the ANPD—National Data Privacy Authority, there was discussion in Brazil whether the person's figure in charge would only fit for lawyers or IT professionals. It was then that in June 2019, MP 869/2018 had in its text the addition of the term "legal-regulatory" as a prerequisite of knowledge for a professional to exercise the role of DPO; because of this, more questions were asked, and such decision would benefit only the legal class.

Seeing this movement, some academic executives and enthusiasts who were part of the technical committees in Brasília raised the flag that it was time for IT professionals to have a representation focused on the topic of privacy and data privacy. At that moment, such a group of academic executives was looking for names that could start a visionary project of national magnitude like this, without political/partisan bias, and aftermarket research, they pointed out the name of Dr Davis Alves to chair such an initiative.

Davis Alves is an IT professional, PhD in the United States, specializing in information security with several international certifications and being one of the first DPOs in Brazil to work abroad, having trained the first Brazilian Data Protection Officers.

In the first half of 2019, Dr Davis Alves accepted the challenge and started to gather people interested in the subject, having his students from the EXIN Privacy & Data Protection Practitioner course from Portal do Training as enthusiasts and future DPOs, who joined as founding members to form the then ANPPD—Associação Brazilian Data Privacy Professionals After the initial meeting with the founding members, Dr Davis Alves sought out big names in the IT area to join the ANPPD steering committee, among them.

Umberto Correia, DPO—IT governance and information security executive of one of the largest Brazilian institutions for the vice presidency.

André Masili, DPO and founder of Grupo Linx/SA for the general secretary; The ANPPD was founded with the mission of bringing together the best Data Privacy Professionals—DPOs in Brazil, promoting technical/scientific knowledge on the subject and representing the class during decision-making in the National Congress involving the LGPD by supporting it. Those with technical bases without partisan and political ends.

In the preliminary provisions of the LGPD, art. 6 reinforces the activities of processing personal data, by the principles of good faith, where information security must use technical and administrative measures capable of protecting this data from unauthorized access and from incidents that cause loss, alteration, communication or dissemination [1].

Another relevant aspect presented by LGPD in its article 11 is the processing of sensitive personal data, which are those that can discriminate against a person, such as racial or ethnic origin and religious conviction, which can only be manipulated in case of prevention of fraud, security of data holder, identification process and registration authentication in electronic systems, exceptions provided for in art. 9 where there is a need to guarantee the protection of this holder's data [1].

The LGPD values data security and confidentiality according to art.46, where processing agents and processors must use security, technical and administrative measures capable of protecting access to unauthorized personal data and from accidental or illegal incidents, such as loss, alteration, and communication, or inappropriate treatment [2].

9.1.3 Artificial Intelligence

Research related to AI started after World War II, and Alan Turing carried out the first work in this area. Since then, much research has been carried out. Defining the concept of artificial intelligence is very difficult. For this reason, Artificial Intelligence was (and remains) a notion that has multiple interpretations, often conflicting or circular.

The difficulty of a clear definition may come from the fact that several human faculties are being reproduced, from the ability to play chess or be involved in computer vision, voice analysis and synthesis, fuzzy logic, artificial neural networks, and many others. Initially, AI aimed to reproduce human thought.

Artificial Intelligence embraced the idea of reproducing human faculties such as creativity, self-improvement, and by use of language Artificial Neural Networks.

The Recurrent Neural Network (RNN) network architecture: The hidden neurons of the recurrent neural network receive the result of the mathematical operation that they performed in the previous time and the data from the previous layer. Because they have this characteristic, these networks can model problems with temporal characteristics, such as the weather forecast given the climate history in a past window. Thus, the RNNs consider a temporal dependency between the input data.

9.1.4 Machine Learning

Neural networks can "learn" to diagnose the most appropriate antibiotic, but they depend on preliminary tests with a structured database evaluated by specialists in actual cases. There are several methods for training the neural network.

The activation function receives the sum of the multiplication of each weight by the respective value of the parameter (or sign) evaluated; this is typically done by the scalar product of the weight matrix by the parameter value matrix. The result of this sum serves as an input parameter for an activation function where a trigger threshold is defined. If the result of the calculation of the activation function exceeds the threshold, a value is propagated to the next neuron ahead. The nonlinear activation function usually gives better results. As well in Fig. 9.4.

Activation function Sigmoid σ (z): (ranging from 0 to + 1): activation notation used in linear function. For each large z, we have σ (z): = 1; for z, small or tending to small, we have σ (z) $\approx$0. In training, the average sigmoid is around 0.5. They are usually used only in the Output layer when a binary is expected.

Tanh-hyperbolic activation function (ranging from -1 to + 1): activation notation used in non-linear function. It is generally used in the output and hidden (learning) layer because in the it is close to 0.0.

Activation function Rectified Linear Unit ReLU: activation notation used in non-linear function. Usually used in the hidden layer (learning).

Activation (look at Fig. 9.1) function Rectified Linear Unit ReLU-Leaky [Loose] (max): activation notation used in non-linear function. Usually used in the hidden layer (learning), so σ (z): = 0.01;

9.2 Bibliographic Review

9.2.1 DLP—Data Loss Prevention

Data Loss Prevention is not new, but it is being used more nowadays, as ready-made tools on the market already have some features of this monitoring standard built-in. Applications such as Office 365 are already able to bring some templates internally standardized for compliance with the GDPR. Even trying to do an analysis for compliance with the regulations, with the lack of depth in the evaluation of certain criteria, data loss can occur [15].

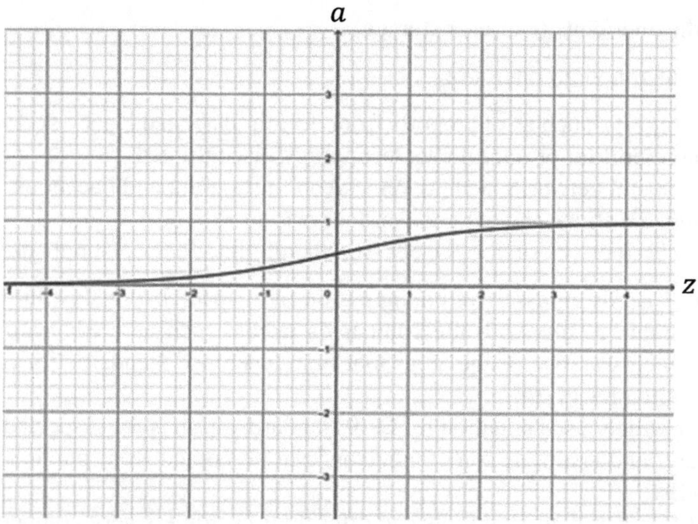

Fig. 9.1 Sigmoid: $a = \frac{1}{1+e^{-z}}$

Other Antivirus tools like Trend Micro, which has the solution called Deep Security for servers only, and another solution called Apex One, which is a SaaS for endpoint security, both have the functionality of reporting DLP logs—Data Loss Prevention, which can be configured according to the company's needs, for example: search for information disclosure of CPF, CNPJ, Invoices, credit card, social security, etc. [15].

Internally, each of these tools works with artificial intelligence algorithms that cross data and monitor patterns pre-formatted by the information security analyst, but directly using classical logic for their conclusions and report presentations so that a manager can make a decision-making process. The Decision whether to continue with the process when a data loss situation is detected.

DLP—Data Loss Prevention helps to delimit the data:

- What are their origins or inputs?
- Where they are stored.
- Within which systems undergo consistency and are transformed into informative reports.
- What are the data outputs?
- And when, how, and where they can be destroyed

Our study delimits the data source into structured and unstructured, where the data that is in:

- From the Network (on the servers and application systems).
- From the Cloud (external storage).
- From Storage (local storage).

- From the endpoints (end-user storage).

While the unstructured ones are:

- Printed reports, such as old reports from decommissioned legacy systems.
- Data on microfilm or microfiche, technology discontinued.
- Data on social networks such as WhatsApp, Instagram, and Facebook.
- Data on USB sticks and external hard drives.
- Data in cloud drives such as OneDrive, SharePoint, Teams, Google drive, etc.

The DLP needs an assertive configuration by a specialist analyst in the area to capture possible results with the company's needs, such as credit card purchase patterns, which was one of the areas that first developed standards for monitoring suspicious situations [16].

Another widely used situation for DLP standardization is for suspected malware, ransomware, and spam, as some indicators can be copied from public safety lists called CVEs.

Not all Incident detection tools can offer their customers very competitive optimization services; however, many need a very refined and carefully protected management, usually by a SIEM—Security Information and Event Monitoring, that is, a tool that manages to unify alerts about malicious activities such as IP scanning, data flow, malicious emails, intrusion attempts, firewall monitoring, review of security policy updates, security patches implemented, but here we will use this concept to create a specific process focused on Data Loss [16].

Additionally, better opportunities to effectively identify redundant and non-ideal elements in organizational and management structures that can be safely identified, changed or removed, bringing the benefit of pre-emption before the data management framework [15].

If there is a demand regarding the outdated business environment, it brings awareness about the need to carry out a complete optimization of the organization and data privacy security management structure preferably using aspects of DLP—Data Loss Prevention to prevent loss or data leakage [17].

The tools usually need some testing and calibration time, so they do not create alerts with excessive false positives but use classic logic [19].

The level of sophistication reached by malicious software requires constant efforts and considerable expenditure of resources to mitigate this practice. The balance of risk appetite between the parties is fragile, as recognized by several authorities in the area. Therefore, one must constantly innovate and be one step ahead of these offenders [18].

9.2.2 Paraconsistent Annotated Evidential Logic $E\tau$

Annotated logics constitute a class of paraconsistent logic. Such logics are related to certain complete lattices, which play an essential role. A knowledge of an expert on an analyzed subject, questions are used to capture opinions that are normalized in logic between 0 and 1, as shown in Fig. 9.3. These values are respectively the favourable evidence that is expressed by the symbol μ and the contrary evidence by λ.

Logic Eτ must follow the process (look at Fig. 9.2) during the application, which can be seen in the figure:

The definition means understanding and creating the proposition to be understood and which should reflect the problem.

The transformation would be treating data from favourable and unfavourable evidence. We usually normalize the answer between the number zero and one to be handled by Logic Eτ.

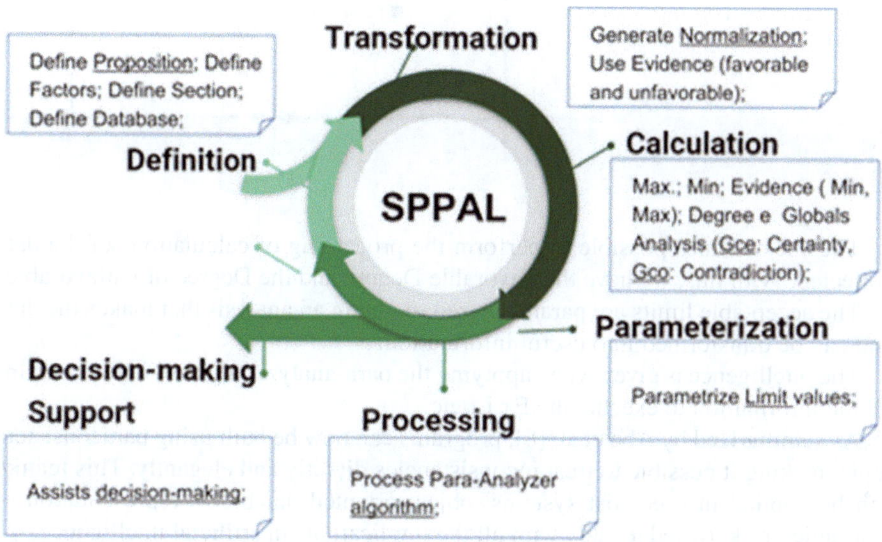

Fig. 9.2 Steps process logic paraconsistent annotated evidential Eτ

Fig. 9.3 Lattice diagram

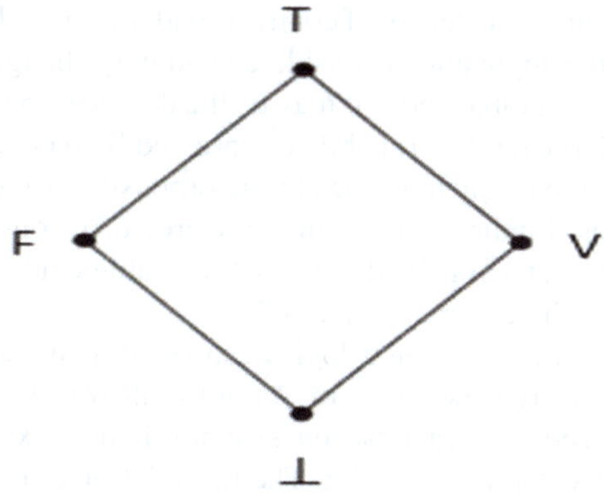

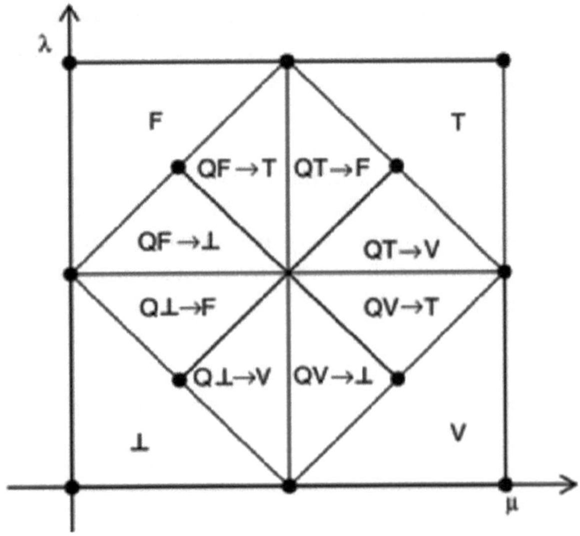

Fig. 9.4 Aspect of the lattice to make decision. *Source* Abe et al. [7]

Thus, it becomes possible to perform the processing of calculations of the data collected. With this, we have the Favorable Degree and the Degree of Unfavorable.

The acceptable limits are parameterized to obtain an analysis that makes the data and can be transformed into useful information.

The intelligence is given when applying the para-analyzer algorithm that contains all the information to execute the Eτ Logic.

As summarized by Abe et al. [8], programs can now be built using paraconsistent logic, making it possible to treat inconsistencies directly and elegantly. This feature can be applied in specialist systems, object-oriented databases, representation of contradictory knowledge, etc., with all the implications in artificial intelligence.

In Abe et al. [8]: "Logic Paraconsistent Annotated Evidential Eτ has an Eτ language and the atomic propositions are of the type p (μ, λ) where p is a proposition and μ, λ ∈ [0, 1]. Intuitively, μ indicates the degree of unfavourable evidence1 of p and λ the degree of contrary evidence of p. The reading of the values μ, λ depends on the applications considered and may change: in fact μ it may be the degree of belief favourable and λ it may be the degree of belief contrary to proposition p; also, μ can indicate the probability expressed by p occurring and λ the improbability expressed by p occurring. The atomic propositions p (μ, λ) of logic Eτ can intuitively be read as: I believe in p with the degree of favourable belief μ and the degree of contrary belief λ, or the degree of favourable evidence of p is μ and the degree of evidence to the contrary of p is λ".

Paraconsistent logics can serve as underlying logic of theories in which A and ¬A (the negation of A) are both true without being trivial [6]. There are many types of paraconsistent systems. In this text, it consider the Paraconsistent Annotated Evidential Logic Eτ. The formulation in Logic Eτ are of the type p (μ, λ), in which p is a proposition and e (μ, λ) ϵ [0, 1] is the real unitary closed interval.

A proposition p(μ, λ) can be read as: "The favorable evidence of p is μ and the unfavorable evidence is λ" [8]. For instance, p (1.0, 0.0) can be read as a true proposition, p (0.0,1.0) as false, p (1.0, 1.0) as inconsistent, p (0.0, 0.0) as paracomplete, and p (0.5, 0.5) as an indefinite proposition [8]. Also we introduce the following concepts: Uncertainty degree: $G_{un\,(\mu,\,\lambda)} = \mu + \lambda—1$ $(0 \le \mu, \lambda \le 1)$ and Certainty degree: $G_{ce(\mu,\,\lambda)} = \mu - \lambda$ $(0 \le \mu, \lambda \le 1)$ [9].

An order relation is defined on [0, 1]: $(\mu_1, \lambda_1) \le (\mu_2, \lambda_2) \leftrightarrow \mu 1 \le \mu 2$ and $\lambda 2 \le \lambda 1$, constituting a lattice that will be symbolized by τ.

With the uncertainty and certainty degrees, we can get the following 12 output states (Table 9.2): extreme and non-extreme states. It is worth observed that this division can be modified according to each application [20].

Para-analyser Algoritm. In this proposed algorithm, there is a set of information obtained, which can sometimes seem contradictory, making it difficult to analyze the scenario for risk analysis. Generally, in such situations, this information is discarded or ignored, that is, they are considered "dirty" of the system, however at best they may even receive different treatment.

Silva Filho, Abe and Torres [7] quote: "However, the contradiction most of the time contains decisive information, as it is like the encounter of two strands of opposing truth values. Therefore, to neglect it is to proceed in an anachronistic way, and that is why we must look for languages that can live with the contradiction without disturbing the other information. As for uncertainty, we must think of a language that can capture the 'maximum' of 'information' of the concept".

In this line of reasoning for the analysis based on Paraconsistent Logic, situations of Inconsistency and Paracompleteness will be considered together with the True and False, represented according to Table 9.1:

The set of these states or objects (τ = {F, V, T, ⊥}) can also be called annotation constants and can be represented using the Hasse diagram as shown in Fig. 9.3.

"The operator about τ é: ~ :|τ| →|τ| that will operate, intuitively, like this:

~ T = T (the 'negation' of an inconsistent proposition is inconsistent).

~ V = F (the 'negation' of a true proposition is false).

~ F = V (the 'negation' of a false proposition is true).

~ ⊥ = ⊥ (the 'negation' of a paracomplete proposition is paracomplete).

Annotated Paraconsistent Logic will be used; this type must be composed of 1, 2 or "n" values.

With the calculations of the values of the axes that make up the representative figure of the lattice, it can be divided or internally delimited into several regions of

Table 9.1 Extreme states fonte: Abe et al. [7]

Extreme states	Symbol
True	V
False	F
Inconsistent	T
Paracomplete	⊥

Table 9.2 Non-extreme states

Non-extreme states	Symbol
Quasi-true tending to Inconsistent	$QV \rightarrow T$
Quasi-true tending to Paracomplete	$QV \rightarrow \perp$
Quasi-false tending to Inconsistent	$QF \rightarrow T$
Quasi-false tending to Paracomplete	$QF \rightarrow \perp$
Quasi-inconsistent tending to True	$QT \rightarrow V$
Quasi-inconsistent tending to False	$QT \rightarrow F$
Quasi-paracomplete tending to True	$Q \perp \rightarrow V$
Quasi-paracomplete tending to False	$Q \perp \rightarrow F$

different sizes and formats, thus obtaining a discretization of the same. From the bounded regions of the lattice, it is possible to relate the resulting logical states, which, in turn, will be obtained by interpolating the Degrees of Certainty Gc and Contradiction Gct. Thus, for each interpolation between the degrees of certainty and contradiction, it is possible to extract information to assist in decision making [20].

The representation of table 9.2 shows a representation of the lattice constructed with values of Degrees of Certainty and Contradiction and sectioned into 12 states. Thus, at the end of the analysis, one of the 12 possible resulting logical states will be obtained as an answer for decision making.

Some additional control values are:

- Vscct = maximum value of uncertainty control = Ftun
- Vscc = maximum value of certainty control = Ftce
- Vicct = minimum value of uncertainty control = −Ftun
- Vicc = minimum value of certainty control = −Ftce

All states are represented in the next Fig. 9.4.

9.2.3 Artificial Intelligence Techniques

In the 1940s, Warren McCulloch and Walter Pitts made the first proposition of an artificial intelligence model, as shown in Fig. 9.10, which used a neuronal structure, suggesting the use of hardware; in this case, they were variable resistors connected to amplifiers, to behave (look at Fig. 9.5) like a human neuron [11].

Its concept was simple, as it was able to model separable linear systems, such as logical operators AND, OR, and NOT. The neuron would input a list of Boolean data (0 or 1), do the sum, and then, in the sequence, do the sum for a triggering function that would return 1 if the sum exceeds the limit and 0 if it fails [11].

Figure 9.10 exemplifies how a neuron would be configured to compute $\times$ 1 AND !X2. Making ! $\times$ 2 an inhibitory input, where there would be only two possible situations: $\times 1 = 0$ and $\times 2 = 0$ and $\times 1 = 1$ and $\times 2 = 0$. Obviously, the expression

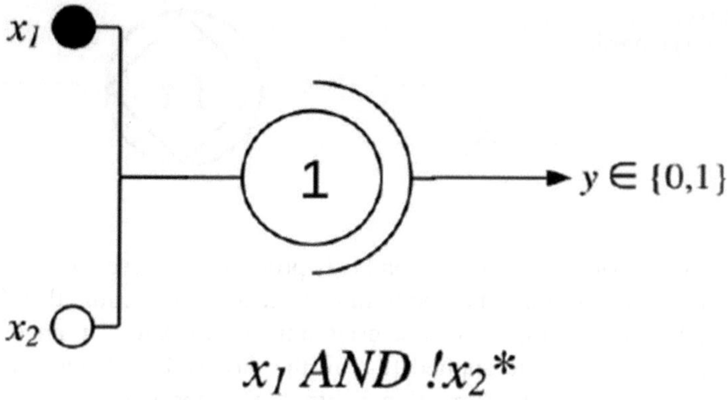

Fig. 9.5 McCulloch-Pitts neuron model

can only evaluate to be true if × 1 = 1, and therefore case 2 is the only valid one [11].

Nowadays, many organizations are looking for AI—Artificial Intelligence solutions to find analogies that lead to optimized containment, being able to use the focus on non-classical logic to circumvent these types of attacks from more offensive groups [20].

The first results of the implementation of methods of convolutional neural networks were carried out with the financial sector, which still prefers not to hire exclusive data analysts because this type of attack attempts that use Artificial Intelligence/Neural Networks, despite having complex problems such as optimization of its cyber structure, according to a study presented by the company Trend Micro. Even the results of the first tests demonstrate the potential of such approaches, which can stimulate companies to think about creating an AI—Artificial Intelligence separate from the internal structure or developing a Machine Learning structure specific to each client [11].

The artificial intelligence technique [12] comes from the 1940s in the studies proposed by Warren McCulloch and Walter Pitts (1943). Some researches are based on neurons' philosophy, knowledge, and function and use propositional logic created by Russell and Whitehead. Another pillar is Turing's theory of computation.

These two researchers proposed a model of artificial neurons, in which each neuron is characterized by being "on" or "off", with the switch to "on" occurring in response the output of a neuron from the layer stimulates the learning of the next layer..

The state of a neuron was considered "equivalent in concrete terms to a proposition that defined its appropriate stimulus". McCulloch and Pitts also suggested that properly defined networks would be able to learn. For example, they showed that a certain network of connected neurons could calculate any computable function and that simple network structures could implement all logical connectives (and, or, not, etc.).

Fig. 9.6 Paraconsistent artificial neuron proposal, author, 2021

In Fig. 9.6, the paraconsistent neuron [1] is proposed to serve the paraconsistent neural network. Another important point must be taken into account that the neuron chooses which characteristic (x) will be used in the network to be trained.

The concept of Eτ Logic applied [24] in the day-to-day of our reality in the face of numerous sources of information, contradiction constantly occupies a space, bringing uncertainties that will culminate in brief or future challenges.

In the case of a system with artificial intelligence [27], neural networks [27], also known as "machine learning" [28], which starts from the study of pattern recognition [20, 21], the appearance of contradiction in reasoning logic is inevitable when we try to reflect human behaviour.

In activities of the medical, hospital, and health segment, it has been observed in the use of analysis of clinical exams, early diagnosis of cancer [26], in politics, in the analysis of lawsuits, and productivity of public security [30], in the measurement of software [25, 29] technical support, in the service of insurance companies, where at least two specialists are involved [27], there will always be different points of view.

In response to the contradiction, we have the Eτ Logic in service in any commercial, scientific segment that can be aggregated with other technologies. An application in six sigma services [31] has been explored both in industry and in services where at least two specialists are involved [27].

The Eτ Logic has been researched in the segment of animal welfare; it has always been in-depth with prediction and the entire agribusiness chain [32].

9.2.4 Data Protection

Brazil has the LGPD—General Data Protection Law in the same standards as the GDPR—General Data Protection Regulation of the European Union; perhaps there is not as much detail as this one and presents the regulatory compliance needs for all Brazilian companies [1, 2].

Despite these needs, many companies in Brazil are still not prepared to comply with the LGPD [1], which can be observed in the research carried out by the class entity called ANPPD - National Association for Data Protection Professionals, which annually surveys its 4 thousand members (look at Fig. 9.7) across the country and concluded in Fig. 9.1 that [3]:

- 84% of companies are in the implementation process.
- 13% are aware of the GDPR compliance needs but do not apply it.

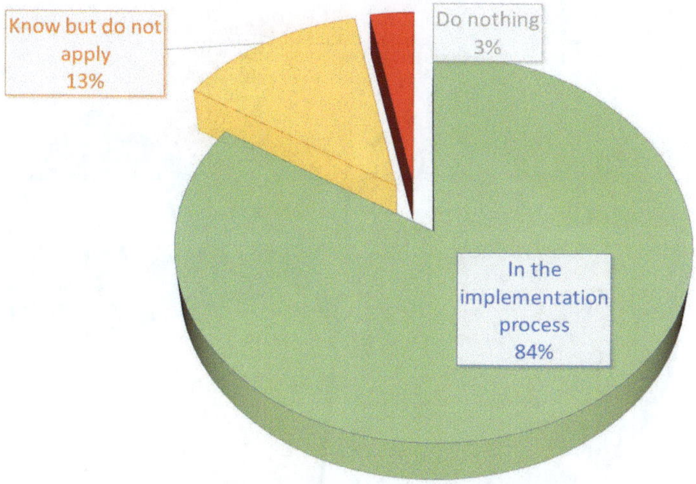

Fig. 9.7 Level of awareness for LGPD compliance in 2022

- 3% do nothing.

Another important point for the development of compliance with the rules is the updating of specialists to mitigate or reduce data loss, this constant concern of both the specialist and the company, as mitigation actions are not only important for adherence to the LGPD. In addition to raising awareness of the entire company together, so that the past problem is not repeated, resulting in fines costs for data leakage incidents [13, 14].

Therefore, the analysis of ANPPD statistics [3, 5] shows this panorama and indicates the concern with the development of adequate training for professionals working with data protection because the better the level of corporate awareness, the lower the operational error due to ignorance of the risk factor (look at Fig. 9.8). In this research, it can be observed that:

- 51% of professionals take specific private courses on LGPD [1].
- 19% participate only in Congresses.
- 16% prefer free courses.
- 14% take a long-term specialization such as an MBA.

The LGPD details in article 5 the data life cycle, which is determined by the following steps [1]:

- Creation that is determined in: collection, production, reception, or data extraction.
- Transport through transmission, distribution, and communication of data.
- Handling focused on: classification, use and modification of data.
- Storage is determined in archiving and storage.
- Discard when data is deleted.

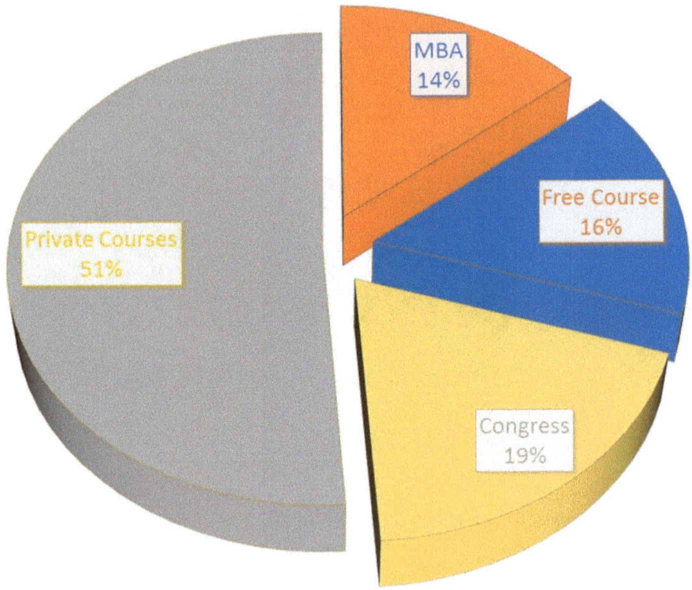

Fig. 9.8 LGPD training

9.3 Minimization of Data Loss

9.3.1 DLP—Data Loss Prevention Using Paraconsistent Annotated Evidential Logic Eτ

When using the concept of DLP – Data Loss Prevention (look at Fig. 9.9), the following granularity was considered to obtain graduation for the types of data loss detected [4]:

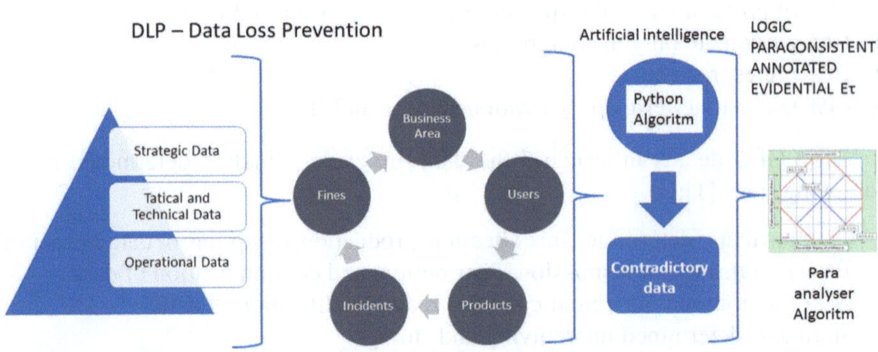

Fig. 9.9 DLP—Data loss prevention using logic paraconsistent annotated evidential logic Eτ

Fig. 9.10 Data source example

	A	B
1	Data	
2	1.1,2.1,3.1,4.1,social-midia	
3	1.1,2.2,3.1,4.2,e-mail	
4	1.1,.2.3,3.2,4.2,storage	
5	1.2,2.4,3.25,4.2,endpoint	
6	1.3,2.5,3.2,4.2,cloud	
7	1.3,2.1,3.1,4.4,cloud	
8	1.3,2.2,3.1,4.3,social-midia	
9	1.3,2.3,3.1,4.2,storage	
10	1.1,2.5,3.1,4.2,endpoint	
11	1.1,2.1,3.1,4.1,social-midia	
12	1.1,2.2,3.1,4.2,e-mail	
13	1.1,.2.3,3.2,4.2,storage	

First level:

- Strategic Data: Loss of confidential data from top-secret projects with greater risk and company exposure in the market.
- Tactical and Technical Data: Loss of confidential information, such as user credentials, system administrators, and authority to transfer large amounts.
- Operational Data: Loss of information on standardization of activities, generating rework or initial mapping needs, and new training for base teams.

The second level [6]:

- Identification of business areas.
- Identification of impacted users.
- Identification of impacted product.
- Identification of incidents with personal data.
- Identification of the cost of the fine due to non-compliance with LGPD

On the third level [11]:

Use the mass of data collected from the financial company for only one month to be analyzed by the Python Algorithm to obtain the report with the answers.

At the fourth level [20], the information from the exceptions presented by the Python Algorithm report is aligned. They are inserted into the Algorithm Para analyzer to identify the percentage of the difference between the responses of the two Algorithms, thus achieving greater assertiveness in making decisions.

DLP tools—in the market have been increasingly sought after by the Company due to LGPD in Brazil, and GDPR in the EU-European Union. Such functionalities can be used as input in Logic Eτ. The Control can be made available through the USB Port like other devices.

The main objective is permanently blocking, monitoring and managing these devices. The systems exploit granular control based on the identification (ID) of the distributor, supplier, customer, product ID, and serial number.

There are several channels that can be used that go through Data Digitization in Motion, being plausible for monitoring and blocking file transfers. All detail is done by inspection of content and context.

Manual or automatic scans are constantly performed in order to exclude sensitive data and ensure DLP. There are cases where data in transit is forcibly encrypted and thus maintains the quality of the DLP technique in the data processing.

According to the Law, it is understood that treatment must be considered in every operation carried out with personal data, such as those referring to the collection, production, reception, classification, use, access, reproduction, transmission, distribution, processing, archiving, storage, elimination, evaluation or control of information, modification, communication, transfer, dissemination or extraction.

Internationally data from outside the national territory and which are not the subject of communication, shared use of data with Brazilian processing agents or object of international transfer of data with another country other than the country of origin, provided that the country of origin provides a degree of protection of personal data adequate to the provisions of the Law.

The use of DLP tools focuses on complying with regulations regarding sensitive personal data: personal data on racial or ethnic origin, religious conviction, political opinion, union affiliation or organization of a religious, philosophical, or political nature, data referring to health or sexual life, genetic or biometric data when linked to a natural person.

Companies should use the DLP concept because they treat data according to the operator and controller precepts. The latter is considered a natural or legal person, governed by public or private law, who is responsible for decisions regarding the processing of personal data. And when the operator is a natural or legal person, public or private, who processes personal data on behalf of the controller;

9.4 Tests

9.4.1 Python Program and Mass Data Results

The mass of data to be tested had the following layout with this legend:

The First Level:

Strategic data = 1.1
Tactical and Technical data = 1.2
Operational data = 1.3

The second level:

- Identification of business areas = 2.1
- Identification of impacted users = 2.2
- Identification of impacted product = 2.3
- Identification of incidents with personal data = 2.4
- Identification of the cost of the fine due to non-compliance with LGPD = 2.5

The third level:

effectively-identified data = 3.1
contradictory data = 3.2

The fourth level:

True = 4.1
False = 4.2
Incomplete = 4.3
Paracomplete = 4.4.

For which each (look at Fig. 9.10) of the captured lines presented a description of the data source.

The mass of data captured presented about 30 pieces of information per day for one month, comprising an amount for analysis of 1,107 data, of which the algorithm verified 60% in Python (look at Fig. 9.11) as effective and 40% as contradictory.

Data considered contradictory were passed on to the para-analyzer algorithm, and a differential of 9% true, 6% incomplete, 10% paracomplete and 15% false was obtained.

However, the interval considered false can pass a more excellent sieve that requires more effective monitoring due to the percentage of occurrences that could be an alert of data leakage in a subtle way.

```python
import csv
from typing import List
from util import normalize_by_feature_scaling
from network import Network
from random import shuffle

if __name__ == "__main__":
    financ_parameters: List[List[float]] = []
    financ_classifications: List[List[float]] = []
    financ_species: List[str] = []
    with open('\\Users\Liliam\Downloads\RNTreinaCompara\RedeNeural_01\financ.csv', mode='r') as financ_fi
        financses: List = list(csv.reader(financ_file))
        shuffle(financses) # get our lines of data in random order
        for financ in financses:
            parameters: List[float] = [float(n) for n in financ[0:4]]
            financ_parameters.append(parameters)
            species: str = financ[4]
            if species == "social-midia":
                financ_classifications.append([1.0, 0.0, 0.0])
            elif species == "e-mail":
                financ_classifications.append([0.0, 1.0, 0.0])
            else:
                financ_classifications.append([0.0, 0.0, 1.0])
            financ_species.append(species)

    normalize_by_feature_scaling(financ_parameters)

    financ_network: Network = Network([4, 6, 3], 0.3)

    def iris_interpret_output(output: List[float]) -> str:
        if max(output) == output[0]:
            return "social-midia"
        elif max(output) == output[1]:
            return "e-mail"
        else:
            return "storage"
```

Fig. 9.11 Python program

9.5 Conclusion

It is concluded that the alignment between the use of Paraconsistent Annotated Evidential Logic Eτ, with the DLP—Data Loss Prevention presents a significant gain for the issue of monitoring and analysis of data loss.

The implementation of this type of tool can increase the company's performance regarding the level of information security, which despised 40% of the data as inconclusive to only 15%, considering that even these intervals can be monitored so that its effectiveness of 100% of the data transited, kept stored and sent by email or analyzed social networks.

Intelligent tools that support companies to mitigate data loss are seen as strategic in corporations.

Both Brazilian and European Union legislation requires the company to respect the data subject and increase the treatment of data surrounding the constant regulation in new businesses.

It is understood that the union of data privacy professionals with the ANPPD brings the possibility of significant advances in the adequacy of the LGPD and an increase in the flow of new business between consolidated countries within the European Union block using the GDPR.

References

1. Brazil. General Personal Data Protection Law. Law no 13.709, Aug. 14th, 2018. Available in: http://www.planalto.gov.br/ccivil_03/_ato2015-2018/2018/lei/l13709.htm. Accessed on: 21 April 2022
2. Lima, Luiz, A.: ANPPD scientific committee statistics: overview of national awareness on LGPD. CNPPD 2022. Available in : https://cnppd.online/. Accessed on: 21 April 2022
3. Bioni, Bruno Ricardo.: Personal data protection: the role and limits of consent. Ed. 1. Vol. single. Rio de Janeiro: Foresee, (2019)
4. Lima, Luiz, A.: ANPPD scientific committee statistics: overview of national awareness on LGPD. CNPPD 2021. Available in : https://cnppd.online/. Accessed on: 21 April 2022
5. Abe, J.M., Nakamatsu, K.: Introduction to annotated Logics—Foundations for paracomplete and paraconsistent reasoning, series title intelligent systems reference library, Volume 88, Publisher Springer International Publishing, Copyright Holder Springer International Publishing Switzerland, eBook ISBN 978–3–319–17912–4, https://doi.org/10.1007/978-3-319-17912-4, Hardcover ISBN 978–3–319–17911–7, Series ISSN 1868–4394, Edition Number 1, 190 pages, (2015)
6. De Carvalho, F.R., Abe, J.M.: Decision making with annotated paraconsistent logic tools. São Paulo. Blucher, pp. 37–47, (2011)
7. Abe, J.M., et al.: Annotated paraconsistent logic evidential Et, pp. 38–39. Comunnicar, Santos, (2011)
8. De Carvalho, F.R., Brunstein, I., Abe, J. M.: Paraconsistent annotated logic in analysis of viability: in approach to product launching. In: Dubois, D.M. (ed.), **718**, pp. 282–291, 2011
9. Dill, R.P., Da Costa Jr., N., Santos, A.A.P.: Corporate profitability analysis: a novel application for paraconsistent logic. Appl. Math. Sci. **8**, (2014)
10. De Lima L.A., Abe J.M., Martinez A.A.G., de Frederico A.C., Nakamatsu K., Santos J.: Process and subprocess studies to implement the paraconsistent artificial neural networks for decision-making. In: Jain V., Patnaik S., Popenţiu Vlădicescu F., Sethi I. (Eds.) Recent Trends in Intelligent Computing, Communication and Devices. Advances in Intelligent Systems nd Computing, Vol 1006. Springer, Singapore. 2019 Print ISBN: 978–981- 13–9405–8; Online Isbn: 978–981–13–9406–5; https://doi.org/10.1007/978-981-13-9406-5_61
11. European Comission Guidelines on Data Protection Officers ('DPOs') (wp243rev.01) 2016 Available in :https://ec.europa.eu/newsroom/article29/item-detail.cfm?item_id=612048 Accessed on: 21 April 2022
12. European Comission Guidelines on Consent under Regulation 2016/679 (wp259rev.01) 2016 Available in :https://ec.europa.eu/newsroom/article29/item-detail.cfm?item_id=623051 Accessed on: 21 April 2022
13. European Comission GDPR—General data protection regulation 2016 Available in :https://gdpr-info.eu/ Accessed on: 21 April 2022
14. Silowash, George J.: KING, Christopher. Insider threat control: Understanding data loss prevention (DLP) and detection by correlating events from multiple sources. Carnegie-Mellon Univ Pittsburgh Pa Software Engineering Inst, (2013)
15. Priest, Graham, Koji Tanaka, Zach Weber.: "Paraconsistent Logic", The stanford encyclopedia of philosophy (Summer 2018 Edition), Edward N. Zalta (ed.), URL = https://plato.stanford.edu/archives/sum2018/entries/logic-paraconsistent/. ISSN: 1095–5054
16. Sikorski, M., Honig, A.: Practical malware analysis: the hands-on guide to dissecting malicious software, No Starch Press (2012)
17. Singh, J., Singh, J.: Challenges of malware analysis: obfuscation techniques, Disponível em: http://50.87.218.19/ijiss/index.php/ijiss/article/view/327. (2018)
18. Akama, S.: Towards paraconsistent engineering, intelligent systems reference library. Springer, Germany (2016)
19. Subrahmanian, V.: On the semantics of quantitative logic programs. In: Proceedings of the 4th IEEE Symposium on Logic Programming, pp. 173–182 (1987)

20. Luiz Antônio de Lima, Liliam Sayuri Sakamoto, Nilson Amado de Souza, Roberto Aures Antonio de Moura, Davis Alves, Claudio Pessoa, José Rogério Poggio Moreira, Jonatas Santos de Souza. "DPO in Brazil from the perspective of the LGPD—General Data Protection Law". EXIN Institute - Ministry of Economic Affairs in the Netherlands. (2020). https://www.exin.com/br-pt/dpo-no-brasil-sob-a-otica-da-lgpd-lei-de-protecao-de-dados/
21. Jonatas S. de Souza, Jair M. Abe, Luiz A. de Lima, Nilson A. de Souza. "The General Law Principles for Protection the Personal Data and their Importance". In: 7th International Conference on Computer Science Engineering and Information Technology (CSEIT 2020—https://arxiv.org/abs/2009.14313), 2020. Computer science & information technology (cs & it), Copenhagen, Denmark. Anais 2020. **10**. p. 109. http://dx.doi.org/https://doi.org/10.5121/CSIT.2020.101110
22. Jonatas, S., De Souza, Jair, M., Abe, Luiz, A., De Lima, Nilson, A., De Souza,: The brazilian law on personal data protection, Int. J. Netw. Secur. & Its Appl. (IJNSA) https://airccse.org/journal/jnsa20_current.html, November 2019, **12**, Number 6, ISSN: 0974–9330; 0975–2307 [Print]. https://aircconline.com/ijnsa/V12N6/12620ijnsa02.pdf
23. De Lima, Luiz A., Abe, Jair M., Kirilo, Caique Z., Da Silva, Jonas P., Nakamatsu, Kazumi.: Using logic concepts in software measurement. Procedia Comput. Sci., **131**, p. 600–607, (2018) http://dx.doi.org/https://doi.org/10.1016/j.procs.2018.04.302
24. Lima, L.A., Abe, J.M., Martinez, a. A.G., Santos, J., Albertini, G., Nakamatsu, K.: The productivity gains achieved in applicability of the prototype AITOD with paraconsistent logic in support in decision-making in project remeasurement". Proceedings of the 9th International Conference of Information and Communication Technology [ICICT-2019] Nanning, Guangxi, China January 11–13, 2019 (http://aivr.org/index.html). Edited by Srikanta Patnaik Volume 154, Pages 1–844 (2019). Procedia Computer Science, pp. 347–353. https://doi.org/10.1016/j.procs.2019.06.050
25. Luiz, A., de Lima, Jair, M., Abe, Angel, A.G., Martinez, Liliam Sayuri Sakamoto, Luigi Pavarini de Lima.: Application of architecture using AI in the training of a set of pixels of the image at aid decision-making diagnostic câncer. In: 25th International Conference on Knowledge Based and Intelligent Information and Engineering Systems (KES2021). 8th–10th September 2021 Szczecin, Poland & Virtual. IS27: Reasoning-based Intelligent Applied Systems: http://KES2021.KESINTERNATIONAL.ORG/CMSISDISPLAY.PHP
26. De Lima, A.W.B., de Lima, L.A., Abe, J.M., Gonçalves, R.F., Alves, D., Nakamatsu, K.: Paraconsistent annotated logic artificial intelligence study in support of manager Decision-Making. In: The 2nd International Conference, 2018, Barcelona. Proceedings of the 2nd International Conference On Business And Information Management—ICBIM' 18. Barcelona, SPAIN: ACM DL, 2018. p. 154- 157. DOI:dx.doi.org/https://doi.org/10.1145/3278252.3278269. https://dl.acm.org/doi/https://doi.org/10.1145/3278252.3278269
27. Luiz A. Lima, Jair M. Abe, Angel A. G. Martinez, Jonatas S. Souza, Flávio A. Bernardini, Nilson A. Souza and Liliam S. Sakamoto.: Study of PANN components in image treatment for medical diagnostic Decision-Making. N.70. The 2nd International Conference on Network Enterprises & Logistics Management—NETLOG 2021. ISSN 2595–0738. http://www.netlogconference.com/papers.html
28. Jonas P. Da Silva, Jair M. Abe, Luiz A. De Lima, Felipe S. David De Oliveira, Kazumi Nakamatsu.: Use of software metrics to scope control in IT projects using paraconsistent logic. Journal WSEAS transactions on computer research. WSEAS transactions on computer research, ISSN/E-ISSN:1991–8755/2415–1521, Volume 6, 2018, Art. #8, pp. 55–59. (2018). https://www.wseas.org/multimedia/journals/computerresearch/2018/a145918-057.php

29. Hugo Gava Insua, Jair M. Abe and Luiz A.de Lima.: Produtividade da Polícia Civil do Estado de São Paulo: uma Análise. IJDR-International Journal of Development Research. ISSN: 2230–9926, Volume:12, Article ID:23962, 4 pages, Research Article. 2022. https://doi.org/10.37118/IJDR.23962.02.2022
30. Kirilo, C.Z., Abe, J.M., Nogueira, M., Nakamatsu, K., Machi Lozano, L.C., de lima L.A.: Evaluation of adherence to the model six sigma using paraconsistent logic, In: 2018 Innovations in Intelligent Systems and Applications (INISTA), Thessaloniki, Greece, 2018, pp. 1–7, INSPEC Accession Number: 18098170, Date Added to IEEE Xplore: September 20 2018, https://doi.org/10.1109/inista.2018.8466287. https://ieeexplore.ieee.org/document/8466287
31. De Alencar Nääs, Irenilza; Duarte da Silva Lima, Nilsa; Franco Gonçalves, Rodrigo; Antonio de Lima, Luiz; Ungaro, Henry; Minoro Abe, Jair.: Lameness prediction in broiler chicken using a machine learning technique". Information Processing In Agriculture, **1**, p. 13, 2020. doi.org/https://doi.org/10.1016/j.inpa.2020.10.003 https://linkinghub.elsevier.com/retrieve/pii/S2214317320302092

Chapter 10
Evaluation of Behavioural Skills Simulating Hiring of Project Manager Applying Paraconsistent Annotated Evidential Logic Eτ

Samira S. Nascimento, Irenilza A. Nääs, Jair Minoro Abe, Luiz R. Forçan, and Cristina C. Oliveira

Contents

Abstract The generation of the knowledge economy adds value to human resources in their behavioural skills favouring the organization. Companies must focus on recruitment and selection processes as their employees build skills in a competitive environment. However, the factors involved in the selection process of project management professionals do not consider behavioural skills as a strategy. The

S. S. Nascimento (✉) · I. A. Nääs · L. R. Forçan · C. C. Oliveira
Paulista University, São Paulo, Brazil
e-mail: sestarisamira@gmail.com

L. R. Forçan
e-mail: luforcan@gmail.com.br

C. C. Oliveira
e-mail: crisolive@ifsp.edu.br

J. M. Abe
Graduate Program in Production Engineering, Paulista University, São Paulo, Brazil
e-mail: jairabe@uol.com.br

J. M. Abe (ed.), *Advances in Applied Logics*, Intelligent Systems Reference Library 243,
https://doi.org/10.1007/978-3-031-35759-6_10

study's objective is to simulate a selection of two candidates for a project manager, applying the Paraconsistent Annotated Evidential Logic Eτ, collaborating with the decision-making process. The materials and methods of this study were based on a hypothetical scenario of changing requirements in a project, experts who evaluate candidates, considering the Behavioural Competency Assessment Instrument. The results indicate overall result values for the first candidate C1 = (0.56, 0.50) "Quasi-Inconsistent tending to True" and the second candidate with values C2 = (0.84, 0.62) "Quasi-Inconsistent tending to False. The study showed favourable collaborative evaluation results, eliminating the process's subjectivity.

Keywords Recruitment and selection process · Behavioral competence · Assessment instrument · Paraconsistent annotated evidential logic Eτ

10.1 Introduction

The changes and expansion of urban centres by external or internal environmental influences force organizations to demand efficient products and services, society demands constant changes, and organizational processes undergo increases to meet market demands.

In this context, there should be a balance between stakeholders in consumer relations, considering the needs of society, market competitiveness, the business environment conducive to the design of innovative ideas and the materialization of processes in products and services [1].

Organizational processes are considered a set of administrative and operational practices involving human factors and their competencies to ensure that the structure and strategy are planned, executed, standardized, and controlled positively [2].

The project has identifiable objectives, consumes management resources, and considers the project lifecycle steps it is closed [3, 4]. The life cycle of a project corresponds to the early phases of the project, the organization and preparation, the execution of the project work and the closure of the project that should reach the expected result.

Project management processes are logically linked by the outputs they produce. Processes can contain overlapping activities that occur throughout the project. Leaving a process can be led to the entry of another process. Or delivery of a project or phase of the project and product.

The project management area is directly related to execution and control, change management, and scope changes. Project requirements need the documented definition of a property or behavior in which a product must unnecessarily meet the project's and management's requirements, ensuring that the project processes are complete within the expected resource range or more similar to the original planning.

By focusing on goals and priorities, project management must respond to strategic changes and adapt to them. The conditions of changes in projects must be observed considering direct and indirect interference. At the same time, stakeholders need to comply with the new decisions, continuous improvement, favour the project management processes, remaining in the management process.

The absence of the set of behavioural competencies of the project manager can be a decisive factor that compromises project management results. The improvement of skills in management balances the needs and productive capacity, besides being a resource to foster organizational strategy.

In the context of the process and design and management, this study defined a hypothetical scenario of changes in design requirements [5], whose objective was to simulate the selection of a project manager with two candidates, evaluating behavioural skills, using an instrument for assessing behavioural skills (IACC) [6], based on Paraconsistent Annotated Evidential Logic Eτ (Logic Eτ) [7].

10.2 Selection Process

Organizations seek to provide better services and products to their customers and potential consumers. Thus, the processes are designed to provide internal balance, with the participation of employees to generate organizational results.

The human recourses process (HR) brings consistency and visibility to people management processes to structure an organization and achieve expected strategic outcomes functionally. In this way, all stakeholders interact. HR processes must assess internal needs following culture and organizational objectives [8].

The recruitment and selection process starts from an internal need of the organization, which requires a descriptive mapping of the candidate profile from the human resources professional. This initial stage of the process assists in choosing the candidate for a given job vacancy. Generally, the recruitment and selection of an applicant must meet the minimum requirements described in the vitae curriculum and possess the specific higher education diploma that meets the technical conditions for a preliminary appointment [9].

The second stage of the process consists of evaluating, classifying, and eliminating candidates. At this stage, specific personal and professional questions are made as a classification method. The final process takes place through the invitation for the candidate to participate in the selection process [10].

This process allows investigating the knowledge, skill, and attitude of candidates who fit the profile desired by organizations. Recruiters should be encouraged to participate in the selection process by strengthening relationships [11].

10.2.1 Interview

An interview is a dialogue between two or more people (interviewer and interviewee). The interviewer asks vital questions to obtain the necessary information from the interviewee. The interviews are orality and direct speech and present with a particular frequency the application of the fundamental processes of communication, and when used correctly, allow the interviewer to withdraw from their questions, elements for reflection and decision-making of the interviewee's answers [12]. Recruiters currently perform the interview method with questions open in temporal order to meet the selection process qualitatively. In their conduct of the interview, candidates respond by considering their experiences in previous work [13].

The historical contextualization method [14] can characterize the degree of depth the evaluator needs to absorb from the candidate for adjustment [15]. The project requirements and management competence are taken into account, extracting the maximum from the construction of situations experienced by the candidate for decision-making [16].

The topical questions, contextualizing the history, are known between the evaluators and the evaluated, being a starting point to structure the evaluation. Within the evaluation context, one can base a hypothetical case of project changes, creating elements for the candidate's responses [17].

The interviews are dynamic and classificatory in which the evaluator proposes the iteration to evaluate the candidate's competence. The intention is to recognize if there is the ability to adapt and if the candidate can think and act beyond individual action, establishing management's organizational and strategic needs [18].

10.2.2 The Project Manager Candidate

As important as assessing the ability to solve problems is to know if the candidate knows how to work with a focus on their knowledge, skills, attitudes and performance of results, in addition to elementary conditions such as time management, planning their day-to-day tasks, aligning them with deadlines and project changes, which ensure that all management processes are consecration [19, 20]. Generally, the recruitment and selection of a manager candidate must meet the minimum requirements described in the vitae curriculum and possess the specific higher education diploma that meets the technical conditions for a preliminary appointment [9].

The project gestor plays a crucial role in controlling and supervising all activities during the project. The project gestor candidate must obtain skills and responsibilities in coordinating activities and a project. This professional will define and monitor the planned actions, delegate tasks and control resources Sithambaram [21], the manager must obtain a set of skills to achieve the integrity of the project scope, and depending on the starting point of the individual and their skills, attitudes, and aspirations must fully meet the context of supporting project management and its life cycle [4].

10.2.3 Simulation

Unlike case studies and comparations, simulations are presented as a method little used in the social sciences to overcome the difficulties of experiments. A simulation is a virtual experiment that requires an operational model to represent all (or part of) a system or its processes. The use of this method is analyzed to legitimize the simulation model, with a hypothesis applying the Logic Eτ, organizing the data aspiring to future events. Thus, the simulation works for both the discovery context and the context of the evidence. This definition brings two essential aspects necessary for simulation: the first is based on tasks, emphasizing what should be done and how to achieve the proposed objective, and the second is the relationship with the simulator itself. The simulation allows several solutions to overcome the difficulties of the analysis and the risks inherent to the activities in question [22, 23].

The tool allows practical experience in a safe environment, followed by guided reflection, generating an impact both academically and in skill and attitudes related to the study, which in turn brought a vital ally to the methodology of an executable simulation [24]. The virtual simulation modality consists of the use by fictitious people, who are part of a hypothetical environment, being expanded to real systems.

The simulation method and the normalization of the hypothetical data obtained the consonance and integration of the Para-analyzer, creating conditions closer to reality and behavioural trends in evaluating candidates for project management.

10.3 Paraconsistent Annotated Evidential Logic Eτ

The algorithm, defined as a process or formula, which follows steps or steps to achieve a specific objective, considers a paraconsistent analysis; the existence of the schematized structured process called the Para-analyzer algorithm works with four external control values defined in the application environment, represented below:

- The true state (V) happens when $\mu = 1{,}0$ e $\lambda = 0{,}0$.
- The false state (F) happens when $\mu = 0{,}0$ e $\lambda = 1{,}0$.
- The inconsistent state (T) happens when $\mu = 1{,}0$ e $\lambda = 1{,}0$.
- The paracomplete state ($\perp$) happens when $\mu = 0{,}0$ e $\lambda = 0{,}0$.

Such values will guide a proposition considered, for example, “true” for decision-making positively and so on. In Fig. 10.1 and Table 10.1 will help introduce the supplementary concepts.

From the conceptual point of view, the Para-analyzer algorithm is composed of a set of information collected through evaluation instruments for decision analysis, delimiting the regions in the reticulate, being a relevant tool for applications of technological interest. In the Para-analyzer algorithm, the areas in the lattice capable of generating the logical states being applied are applied to the extreme and non-extreme states. Each state receives the denomination that reflects the trends in Table 10.1 [26].

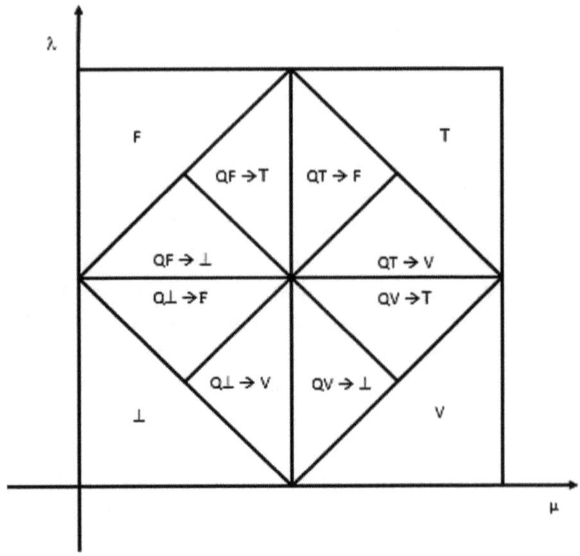

Fig. 10.1 Representation of the lattice τ. *Source* Abe [25]

Table 10.1 Symbolization of logical states

Extreme states	symbol	Non-Extreme states	Symbol
True	V	Quasi-true tending to Inconsistent	$Q_V \rightarrow T$
False	F	Quasi-true tending to Paracomplete	$Q_V \rightarrow \perp$
Inconsistent	T	Quasi-false tending to Inconsistent	$Q_F \rightarrow T$
Paracomplete	⊥	Quasi-false tending to Paracomplete	$Q_F \rightarrow \perp$
		Quasi-inconsistent tending to True	$Q_T \rightarrow V$
		Quasi-inconsistent tending to False	$Q_T \rightarrow F$
		Quasi-paracomplete tending to True	$Q_\perp \rightarrow V$
		Quasi-paracomplete tending to False	$Q_\perp \rightarrow F$

Source Abe [25]

10.4 Method

This section presents the methodological procedures applied in this study to achieve the simulation results by evaluating two candidates for project management and their behavioural competencies. The research shows an exploratory objective, to provide greater familiarity with the problem, intending to make it more explicit or to construct hypotheses by stimulating the understanding of the evaluation of candidates. The research objective was established in the methodological framework, the type of research; the collection instrument; the group's evaluation process, and the analysis of the data and expected results for each candidate.

An orderly series of open questions was contained for the data collection instrument. The candidates must answer that to evaluate their behavioural skills, and the questions are based on the IACC [6]. The questions are divided into two dimensions: μ represents favourable evidence, and λ represents unfavourable evidence, considering the nine behavioural competencies applying the Logic Eτ, as shown in Table 10.2. In addition, the interview aims to capture the experience of candidates associating a hypothetical situation of the project change, contextualizing the main structure of the evaluation.

The simulation was divided into two stages. In the first stage, the candidate receives the eighteen questions in electronic form of the Microsoft forms system®, relating them to behavioural skills and the hypothetical situation of change in the project. In each question, the candidate is asked to frame his decision by describing his actions with open and conclusive answers.

In the second stage of the process, the evaluators receive the questionnaire electronically from the candidates analyzing the decisions made and classifying the answers with values between [0; 100]. The rating values of the evaluators are entered in the Para-analyzer by applying Logic Eτ.

The methodological stages of the evaluation process can be observed in the representation of Fig. 10.2.

10.4.1 The Hypothetical Scenario

A narrative was established with a central issue of the problem for elaborating the hypothetical scenario: the change of project requirement. The main theme of the problem was based on the ISO Standard for establishing stakeholders and their decisions involvement. A contextual analysis was proposed to the project manager for decision-making regarding emetics.

The elaboration of the form contextualizes a hypothetical scenario in changing project to obtain the central understanding of the questions directing the candidate in the answers. The text has been reproduced as described below:

> *A construction company seeks a project management professional to manage its work. According to the planning premises, the project requires a harmonious symbiosis between project managers, builder and client. The integrated process involves the full range of professionals involved in major design decisions, from the design stages to the completion. In order, there are changes, opportunities and solutions or individual points of view that can result in project management with executive advantages and disadvantages. For the project in question, the initial scope was approved with the apparent electrical installations working package, ensuring all the premises of the project and the layout desired by the client. Recital a Norma ISO Sustainability in Building Construction 21931* [27]*, which defines a method for evaluating the environmental performance of construction works, the client requested the immediate change of the project to embed the entire installation in the structure, maintaining access to the infrastructure for maintenance in the networks.*

With this hypothetical context, the project manager makes the initial evaluation of the changes and answers the questions considering the problem:

Table 10.2 Questions to the candidate

Questions to candidates (μ)	Questions to candidates (λ)
1. How can it be adapted and changes requested by the customer more flexible?	10. Assume the team submitted changes primarily at the cost of the design review and were not approved by the customer. How would the team go about negotiating and exposing the actual changes?
2. Given the changes requested by the client, what is the immediate solution to present assertively and appropriately new opportunities in the project?	11. Assume that the team identified new solutions and opportunities regarding design changes, and the customer rated the solutions as failures. What is the decision to follow for steps to resolve customer considerations?
3. What would be the satisfactory solution to the client to mitigate the risks without causing significant conflicts between the parties?	12. What immediate solution should be presented if the client did not accept any trades?
4. The design team disagrees with the actual changes. What is the collective effort to resolve the team's uncertainties, even in the face of customer requests?	13. Given the changes requested by the customer, the project team agrees to the actual changes. What would a collective effort be made to convince the team that the changes are uncertain?
5. Regarding design requirements, stakeholders sign the project start date. What is the best strategy to achieve tangible and valid results for the situation presented by the client?	14. Stakeholders set the project start date. What is the best strategy to achieve tangible and valid results, considering that the client has imposed the changes in the project?
6. Explain what it is like to be a leader and influence in the face of customer-imposed changes to achieve project management goals?	15. The team is discouraged from responding to changes imposed by the customer. How can the team be influenced for changes?
7. What are the best practices and strategies in arguing in a team to face the changes requested by the client?	16. What are good practices and strategies in argumentation with stakeholders (all involved in the project), given the changes imposed by the client?
8. What decision should be made in the face of changes in facilities to seek alternatives and ensure the best results in the project?	17. Faced with the changes imposed by the client, he does not know what the best decisions are. What alternatives should be presented to the client, ensuring the best tangible and measurable result?
9. What is the main risk in the project's outcome if the client has requested changes that are not fulfilled?	18. What are the main risks in the project results if the client's imposed changes are not met?

Source Authors

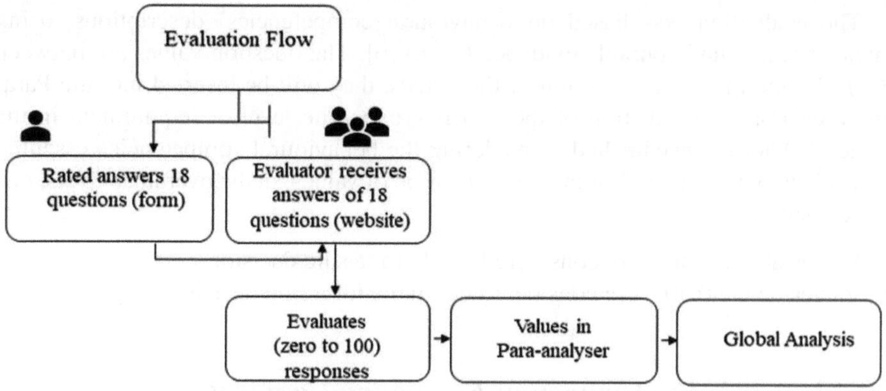

Fig. 10.2 Stage of the evaluation process. *Source* Authors

Judging by what is seen in the change requested by the client, what are their decisions and definitions for this request submitted? Consider all the issues following decision making.

10.4.2 Expert Groups

The application of the Logic Eτ is considered the primary tool for dealing with decision-making problems involving a small number of evaluators. Evaluator groups often differ in knowledge, technical skills, and experience. In this sense, the group of experts will be fully qualified to contribute equally to the resolution of the process and decide to choose the candidate. The professionals selected for this evaluation come from the groups G1 human resources (HR) and the G2 general construction management (GCM), direction construction (CD), according to Fig. 10.3.

Fig. 10.3 Evaluator groups. *Source* do Nascimento et al. [6]

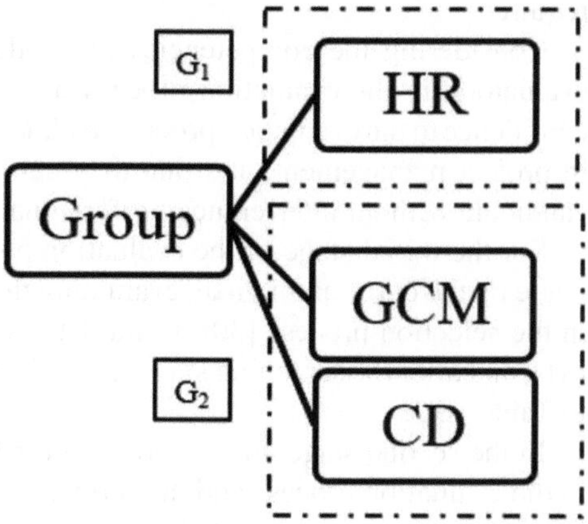

The evaluation was based on behavioural competencies' descriptions, using favourable and unfavourable evidence (μ, λ) [6]. The question values are between [0; 100], and after the collection of the values, they will be inserted into the Para-analyzer. For the evaluation of the expert groups, the level of requirement in the value of 0.50 was established, considering the behavioural competencies essential for project management. For the normalization of values, in the overall analysis, one can consider:

- Values above 0.50—are considered stable for a safe decision
- Values below 0.50—are considered dubitable for a safe decision

10.4.3 Logic Eτ Application Evaluating Candidate

The IACC was used to assist in the collection and analysis of data from the process of evaluation of behavioural competencies of candidates and project management. The instruments have been increasingly used as auxiliaries in evaluations in different aspects of management processes. The common use of qualitative and quantitative research allows us to collect information on what can be achieved in isolation from an evaluation.

For the use of the tool, it is recommended to have an initial understanding of the tool methodology and that, according to the normative premises based on ISO 10015—Guidelines for competence management and human development [28], as competencies and development are being interrelated human, being a requirement in human development. Therefore, all human capital must be developed and trained to understand its processes and organizational structure, and produce products or services according to the strategy, with functional responsibilities to achieving customer service and satisfaction. Furthermore, according to Wandersman [29], everyone involved in the organizational structure, whether internal or external, should receive instruction and training to ensure the quality and management practices required.

Considering the competencies of all those involved in the process, especially evaluators, in the evaluation process of candidates, it is necessary that they offer confidence to the evaluation process, by knowledge, techniques the time of experience in project management, and that they can make decisions in the evaluation of the candidate without interference professionals, ensuring the best choice.

For the initial stage of the evaluation process of a candidate, a prior selection is made by the curriculum vitae, evaluating the potential of the candidate to participate in the selection process [30]. Candidates will be submitted to a distance interview [31], and after the screening stage, candidates receive questions by e-mail, according to Table 10.2.

In the second stage, the experts receive by e-mail the responses of the candidate in the evaluation process and numerically classify the responses on a scale between [0, 100]. The values considered by the experts allow parameterization in the Para-analyzer by applying Logic Eτ.

Table 10.3 Evaluation of candidates

Competence	HR				GCM				CD			
	C_1		C_2		C_1		C_2		C_1		C_2	
	Q 01 a Q 09	Q 10 a Q 18	Q 01 a Q 09	Q 10 a Q 18	Q 01 a Q 09	Q 10 a Q 18	Q 01 a Q 09	Q 10 a Q 18	Q 01 a Q 09	Q 10 a Q 18	Q 01 a Q 09	Q 10 a Q 18
	μ	λ	μ	λ	μ	λ	μ	λ	μ	λ	μ	λ
Adaptation and flexibility	0,90	0,61	1,00	0,20	0,79	0,83	0,64	0,93	0,98	0,29	0,48	0,36
Undertake	0,30	0,80	1,00	0,90	0,80	0,78	0,68	1,00	0,84	0,32	0,32	0,74
Negotiating	0,90	0,17	0,90	0,75	0,72	0,55	0,68	1,00	0,89	0,31	0,40	0,70
Teamwork	0,51	0,15	0,90	1,00	0,70	0,69	0,69	1,00	0,87	0,29	0,87	0,79
Planning	0,27	0,85	0,91	0,15	0,69	0,65	1,00	1,00	0,79	0,24	0,51	0,71
Leadership	0,30	0,60	0,91	0,85	0,53	1,00	0,88	0,80	0,79	0,31	0,81	0,76
Communication	0,91	0,20	0,80	0,85	0,80	0,69	1,00	1,00	0,83	0,16	0,69	0,74
Decision making	0,91	0,40	0,80	0,10	1,00	0,65	1,00	0,92	0,88	0,24	0,81	0,80
Results	0,11	0,90	0,10	0,90	0,64	1,00	0,85	0,64	0,83	0,74	0,43	0,68

Source The Authors

In the third stage, the para-analyzer was used, and the values of the evaluations were randomly generated, allowing the treatment of uncertainties and inconsistencies [7]. The values are between [0, 1], as shown in Table 10.3.

10.5 Result

Comparative analysis is a research method that relies on gathering information that involves parity between two or more processes, documents, datasets, or other objects to obtain justifications for explaining differences or similarities. The comparative analysis between two candidates evaluating behavioural skills was parameterized using the Para-analyzer; the value database is demonstrated in Table 10.3.

The results of the evaluations were compared between candidates C_1 and C_2, with the idea of relating the behavioural skills of two candidates in different terms, applying logic $E\tau$ to reinforce and emphasize the indicators and trends of each candidate. The factors represented by the symbol "triangle" defines as resulting behavioural skills, following the classification according to Fig. 10.1.

The result of the analysis indicates that candidate C_1 presents characteristic "Truth" in the negotiation and communication skills, in the competencies adaptation and flexibilization, undertake and results in present "Inconsistent", indicating dubiety or indeterminacy for a safe decision. The competence teamwork indicates "Quasi-Paracomplete tending to True", which expresses an uncertainty with viable

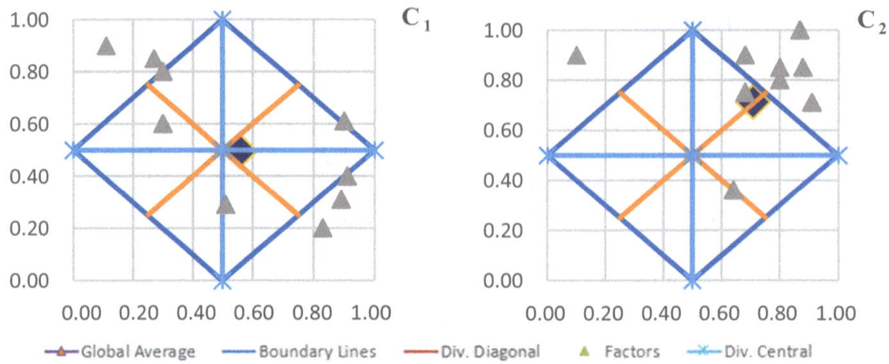

Fig. 10.4 Candidates Para-analyzer results C_1 e C_2. *Source* The Authors

possibilities in meeting the competence, indicating duties or for a safe decision. The planning and leadership skills show a trend "Quasi-Inconsistent tending to True" and "Quasi-Inconsistent tending to False" expressing indecision, indicating dubiety for a safe decision. The decision-making competence means a trend Quasi-True tending to Inconsistent, says uncertainty, dubitable value for a safe decision, as shown in Fig. 10.4.

The results for candidate C_2, present the characteristic "False" in the competence to undertake, contrary to a safe decision. In the competency, teamwork, leadership, communication, and decision-making present "Inconsistent", indicating dubiety or indeterminist for a safe decision. The competence adaptation and flexibilization indicate "Quasi-Paracomplete tending to True", which expresses an uncertainty with viable possibilities in meeting the competence, indicating dubiety for a safe decision. Planning and results indicate a trend "Quasi-False tending to Inconsistent" for negotiating competencies, expressing indecision, and dubitable value for a safe decision, as shown in Fig. 10.4.

Global Comparative Analysis

The Global Comparative Analysis (GCA) comparative para-analyzer algorithm for the candidate (C_1) GCA = (0,56; 0,50) is in the region of the Non-Extreme States "Quasi-Inconsistent tending to True" in QUPC. For the candidate (C_2), GCA = (0,84; 0,62) is in the region of the Neighboring Non-Extreme States "Quasi-Inconsistent a False" and "Quasi-Inconsistent tending to True" in QUPC, as demonstrated in Table 10.4.

In comparing the simulation results using the para-analyzer according to Table 10.4, it was possible to obtain the positioning of behavioural skills and trends by applying the logic Eτ criteria for each condition.

The results found expected by the hypothesis of this study. One can conclude; that in case there is a view of the results of the candidates and the expected results for the management of the project, it will be necessary to examine where the reality and

Table 10.4 Result of logical states

Competence	C_1	C_2
Adaptation and flexibility	**Inconsistent**	Quasi-true tending to Paracomplete
Undertake	**FALSE**	**Inconsistent**
Negotiating	**TRUE**	Quasi-inconsistente tending to false
Teamwork	Quasi-true tending to Paracomplete	**Inconsistent**
Planning	**FALSE**	**Inconsistent**
Leadership	Quasi-false tending to Paracomplete	**Inconsistent**
Communication	**TRUE**	**Inconsistent**
Decision Making	**TRUE**	**Inconsistent**
Results	**FALSE**	**FALSE**

Source The Authors

conformity of the actions come from, elaborating a comparative analysis as suggested in Table 10.4.

The new hypotheses can be constructed to contribute to decision-making scenarios evaluating candidates. Conflicts are part of the evaluation process and may present a biased or contradictory result.

The Para-analyzer algorithm was used to enter the expert's judgment when evaluating the candidates to eliminate the contradiction. This instrument is based on the Logic Eτ, which through the normalization of quantitative opinion, enables or reaches a result and a decision, consolidating a collective and collaborative evaluation. In an actual situation, the evaluation of candidates without a consolidated logical method will remain with subjective decisions or even contradictory.

This research demonstrates that in a simulation environment, hypothetical, it is possible to apply the Paraconsistent Annotated Evidential Logic Eτ in an analysis of behavioural skills.

10.6 Discussion

The results presented in this simulation allowed us to make an acritical analysis to verify if the established requirements of a project can be achieved and the improvements that can add value to the project. Because of the results obtained by comparing the candidates for project management by applying Logic Eτ, the analysis tool's decision-making process helps assess the behavioural competence relationships between two project management candidates. Considering that the basic principles of management are directly linked to the essential competencies of a manager, depending on the results of behavioural skills, does not enable management, representing a risk to the project. The ideal for a safe decision is the characteristics of behavioural skills to present in the Extreme State the "True".

Thus, processing the para-analyzer allowed the evaluation of qualitative criteria and submission in logical values and their trends, classifying the behavioural competencies of each candidate evaluated. For a final representation of the simulation results, Table 10.4 compares the two candidates.

10.7 Conclusion

Simulation research has become increasingly widespread to evaluate behaviour and possible outcomes and scenarios. Therefore, the objective of this study was simulated by a group of evaluators investigating the relevance of evaluating two candidates for project management. The starting point for this simulation is using IACC to generate random data that is applied to a secondary analyzer based on Eτ Logic aiding in the results, comparing two candidates and as supposed references for decision making.

When simulating the evaluations of project management candidates, global results and critical analyses contribute to the body of knowledge; the application of Eτ Logic can be used to evaluate other scenarios and corporate activity. The work demonstrated that through a hypothetical simulation, inserting the data in the para-analyzer, it was possible to obtain qualitative and decisive results. When specialists have a high degree of professional maturity, it is possible to form several groups to evaluate the influences and contradictions in a contract.

Acknowledgements This study was financed in part by the Coordenação de Aperfeiçoamento de Pessoal de Nível Superior—Brasil (CAPES)—Finance Code 001—Process number: 88887.663535/2022-00.

References

1. Carson, P.P., Carson, K.D., Knouse e, S.B., Roe, C.W.: Balance theory applied to service quality: A focus on the organization, provider, and consumer triad. J. Bus. Psychol. **12**, p. 99–120 (1997)
2. Kirchmer, M., Scarsig e, M., Frantz, P., BPM CBOK : version 4.0, springfield, Ill: Guide to the business process management common body of knowledge ABPMP (2019)
3. Kerzner, H.: Strategic planning for project management using a project management maturity model, John Wiley & Sons (2002)
4. PMI®: A guide to the project management body of knowledge (PMBOK® Guide) Sixth Edition. Proj. Manag. Inst. (2018)
5. Popper, K.: The poverty of historicism, III. Economica **12**, 69–89 (1945)
6. do Nascimento, S.S., de Alencar Nääs, I., Abe, J.M., de Oliveira e, C.C., Forçan, L.R.: Instrumento de avaliação de competências aplicando a lógica paraconsistente anotada evidencial Et. Res., Soc. Dev., **10**, p. e7610413444 (March 2021)

7. Abe, J.M.: Sistemas baseados em inteligência paraconsistente: Novas tendências nas aplicações da paraconsistência, Springer, Ed. (2015)
8. Kianto, A., Sáenz e, J., Aramburu, N.: Knowledge-based human resource management practices, intellectual capital and innovation. J. Bus. Res. **81**, p. 11–20 (2017)
9. Galdino e, G.M., Gotway, M.: The digital curriculum vitae. J. Am. Coll.E Radiol. **2**, p. 183–188 (2005)
10. Correa e, H., Craft, J.: Input–output analysis for organizational human resources management. Omega, **27**, p. 87–99 (February 1999)
11. Williamson, I.O., Lepak e D.P., King, J.: The effect of company recruitment web site orientation on individuals' perceptions of organizational attractiveness. J. Vocat. Behav, **63**, p. 242–263 (2003)
12. Corcoran e, J., Burnett, R.: Resident Candidate Interviews," Plast. Reconstr. Surg—Glob. Open, **4**, p. e770, (June 2016)
13. Jackson e, K.M., Trochim, W.M.K.: Concept mapping as an alternative approach for the analysis of open-ended survey responses. Organ. Res. Methods, **5**, p. 307–336, (2002)
14. Rhodes e, L., Dawson, R.: Lessons learned from lessons learned. Knowl. Process. Manag. **20**, p. 154–160 (2013)
15. Hunter, M.G.: Qualitative interview techniques, ECRM, Dublin, Ireland, (2006)
16. Newcomer, K.E., Hatry e H.P., Wholey, J. S.: Conducting semi-structured interviews, Handb. Pract. Program Eval. 492, (2015)
17. Becker, H., Berger, P., Luckmann, T., Burawoy, M., Gans, H., Gerson, K., Gerson, K., Gerson, K., Glaser, B., Strauss e, A.: others, Observation and interviewing: Options and choices in qualitative research. Qual. Res. Action, **6**, p. 200–224 (2002)
18. Back e, K.W., Gergen, K.J.: Idea orientation and ingratiation in the interview: a dynamic model of response bias. (1943)
19. McLagan, P.A.: Competencies: The next generation. Training & development **51**, 40–48 (1997)
20. Englund e R., Bucero, A.: The complete project manager: Integrating people, organizational, and technical skills, Berrett-Koehler Publishers, (2019)
21. Sithambaram, J., Nasir e, M.H.N.B.M., Ahmad, R.: Issues and challenges impacting the successful management of agile-hybrid projects: A grounded theory approach, Int. J. Proj. Manag. **39**, p. 474–495 (2021)
22. Kleiboer, M.: Simulation methodology for crisis management support. Int. J. Proj. Manag. **5**, 198–206 (1997)
23. Zikmund, W.G., Babin, B.J., Carr e, J.C., Griffin, M.: Business research methods, Ninth Edition ed., C. Learning, Ed., South-Western College Pub, (2010)
24. Bruyne, P., Herman e J., Schoutheete, M.: Dinâmica da pesquisa em ciências sociais. P. 84, Editora Francisco Alves, Rio de Janeiro, RJ, Brasil, (1977)
25. Abe, J.M.: Paraconsistent intelligent-based systems: New trends in the applications of paraconsistency, **94**, Springer, (2015)
26. Akama e others, S.: Towards paraconsistent engineering, Springer, (2016)
27. International Organization for Standardization, ISO 21931–1 Framework for methods of assessment of the environmental performance of construction works, ISO, EUA, (2010)
28. International Organization for Standardization, ISO 10015—Quality management — Guidelines for competence management and people development, (2019)
29. Wandersman, A., Chien e, V.H., Katz, J.: Erratum to: Toward an Evidence-Based system for innovation support for implementing innovations with quality: tools, training, technical assistance, and quality assurance/quality improvement. Am. J. Community Psychol, **50**, p 460–461 (June 2012)
30. Hendry, P.: Engendering curriculum history, Routledge, (2011)
31. Marcón, O.A.: Las entrevistas a distancia en Trabajo Social Forense: reflexiones teórico-prácticas. Itiner. Trab. Soc., p 87–94, (January 2021)

Chapter 11
A Paraconsistent Decision-Making Method

Fábio Romeu de Carvalho

Contents

Abstract Considering that decisions in organizations, in most cases, are made based on the decision maker's intuition and experience only, we believe that it would be convenient to find a method for them to be taken based on real data and on a scientific basis. Considering that the decisions are influenced by many factors and that, normally, the data referring to the factors present inconsistencies and uncertainties, we decided that an appropriate logic to treat these data would be the paraconsistent annotated evidential logic Eτ, which deals with data with these characteristics without become trivial or to collapse. Given these considerations, we developed a decision method, which was called Paraconsistent Decision-Making Method, PDM, based on the tools of Eτ logic. This is what we present in this work. For clarity, an example of application of the method is presented in a common decision in higher education: the convenience to open (or not) a higher course in a given region.

F. R. de Carvalho (✉)
Universidade Paulista—UNIP, Programa de Doutorado em Engenharia de Produção, Rua Dr. Bacelar, 1.212, São Paulo, SP, CEP 04026-002, Brasil
e-mail: fabioromeu@unip.br

J. M. Abe (ed.), *Advances in Applied Logics*, Intelligent Systems Reference Library 243,
https://doi.org/10.1007/978-3-031-35759-6_11

11.1 Introduction

By reviewing the literature in the area, we found that the problem related to decision making remains alive and far from reaching a definitive solution. We also find that, in most companies with which we have contact, decisions are made with many difficulties. These result from the existence of a very large number of factors that influence them. In view of this, in most cases, decisions are made based on some inaccurate data, cataloged in a sparse and non-methodological way, and on the sensitivity and experience of the responsible leader [9].

Based on these findings, we thought about the possibility of creating a method, with scientific basis and based on logical criteria, to assist in these decision-making aspects. That is what we did and what we are presenting in this work.

A paraconsistent logic allows manipulating concepts such as uncertainty and inconsistency, without becoming trivial. It accepts "it could be yes" or "it could be no", each of which with a degree of possibility. Due to these characteristics, we believe that there was a possibility, using the tools of this type of logic, to create a process to aid decision making. We will use the evidenced annotated paraconsistent logic call Eτ.

A first work towards the complete systematization of the annotated logic was by N.C.A. da Costa, V.S. Subrahmanian and C. Vago, published in 1991 [4], which makes the first syntactic and semantic study of such logics, dealing with its correction and completeness at the propositional level. Subsequently, Da Costa, Subrahmanian and Abe extended the annotated logic to the quantificational level. This fact leads to another application of these logics, as shown in the work [5].

More recently, J. M. Abe and other researchers have developed applications for paraconsistent logic and annotated paraconsistent logic, with emphasis on applications in computer science, artificial intelligence, robotics, and other domains.

In classical logic, it can be proved that if a theory presents a contradiction, that is, if a formula A is theorem of the theory and its negation $\neg$ A so is, then any formula B of the theory can be demonstrated. This means that a contradiction trivializes a theory that has the classical as its underlying logic. In paraconsistent logic, this does not happen, because, despite being inconsistent, it is non-trivial.

Intuitively, in logic Eτ, what is done is to assign an annotation (***a***; ***b***), with ***a*** and ***b*** belonging to the real unitary interval [0, 1], to each elementary proposition **p**, in such way that ***a*** translates the degree of evidence favorable to **p** and ***b***, the degree of evidence contrary to **p**. Each pair (***a***; ***b***) constitutes a logical state. The following extreme situations stand out:

(1; 0) which intuitively represents total favorable evidence and no evidence contrary to **p** (translates the logical state that is called **truth** and represented by **V**);

(0; 1) which intuitively represents no favorable evidence and total evidence contrary to **p** (translates the logical state that is called **falsity** and represented by **F**);

(1; 1) which means, simultaneously, total favorable evidence and total evidence contrary to **p** (translates the logical state that is called **inconsistency** and represented by **T**), and.

(0; 0) which indicates no favorable evidence and no evidence contrary to **p** (translates the logical state which is called **paracompleteness** and represented by ⊥).

11.2 The Unitary Square of the Cartesian Plane (USCP)

The set of all annotations (***a***; ***b***), which translate the degrees of favorable and contrary evidence that are attributed to the propositions, can be represented by a unitary square (Fig. 11.1), called the Unitary Square of Cartesian Plane (USCP) [5].

This square is nothing more than the closed region [0, 1] X [0, 1] of the Cartesian plane.

We define degree of uncertainty: $G = \boldsymbol{a} + \boldsymbol{b} - 1$, and degree of certainty: $H = \boldsymbol{a} - \boldsymbol{b}$.

From these definitions, we conclude that: $-1 \leq G \leq 1$ and $-1 \leq H \leq 1$.

USCP can be divided in different ways. A convenient way to divide it is into twelve regions, using, in addition to the PUL (AB) and PDL (CD) lines, for example, the following boundary lines:

|G| = 0.60 → MN: $G = \boldsymbol{a} + \boldsymbol{b} - 1 = -0.60$ (completeness limit line) and.
RS: $G = \boldsymbol{a} + \boldsymbol{b} - 1 = +0.60$ (inconsistency limit line).
|H| = 0.60 → TU: $H = \boldsymbol{a} - \boldsymbol{b} = -0.60$ (falsity limit line) and.
PQ: $H = \boldsymbol{a} - \boldsymbol{b} = +0.60$ (truth limit line).

In this case, the USCP division is represented as in Fig. 11.2.

With this division we can highlight four extreme regions and a central region.

AMN Region: $-1 \leq G \leq -0.60$ (paracompleteness region).

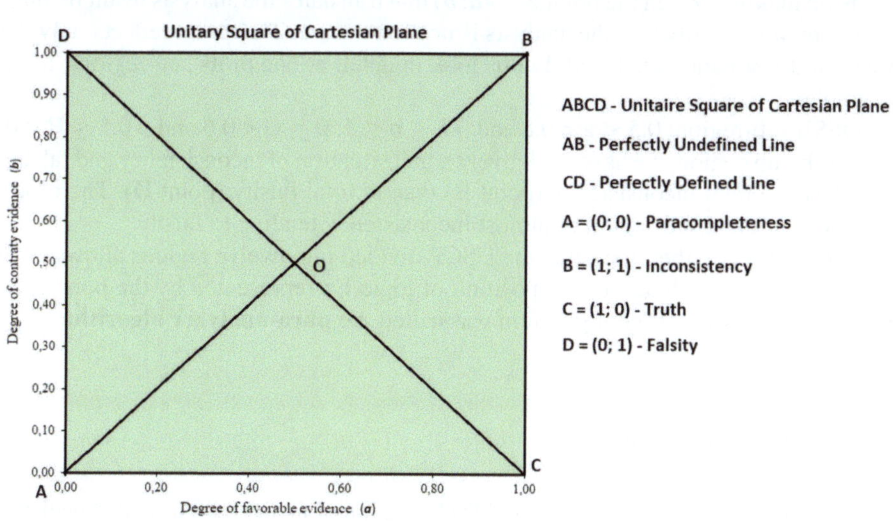

Fig. 11.1 The unitary square of cartesian plane

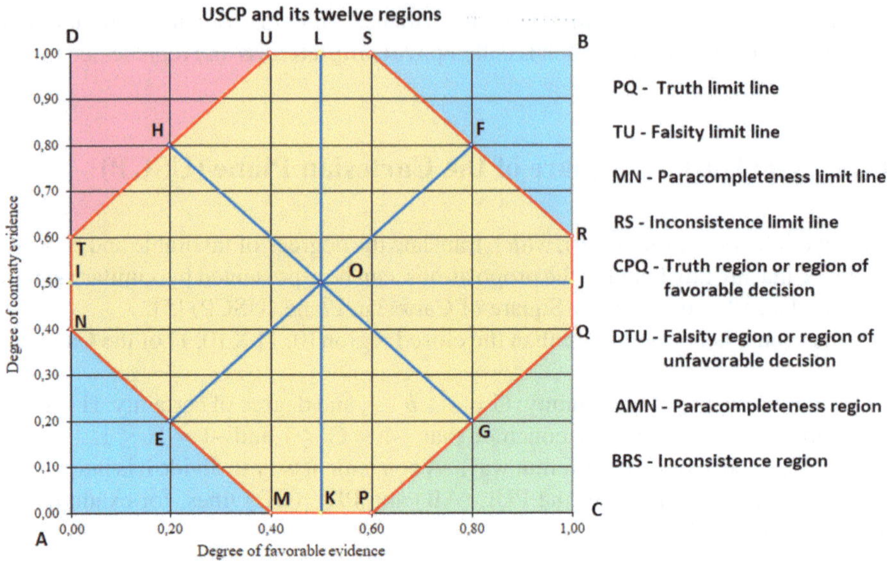

Fig. 11.2 The twelve regions of the unitary square of cartesian plane

BRS Region: $0.60 \leq G \leq 1$ (inconsistency region).

CPQ Region: $0.60 \leq H \leq 1$ (truth region).

DTU Region: $-1 \leq H \leq -0.60$ (falsity region).

MNTUSRQP Region: $-0.60 < G < 0.60$ and $-0.60 < H < 0.60$ (central region).

This is the region that we will call "non-conclusive region", as it does not allow decision making. When the point $X = (\boldsymbol{a}; \boldsymbol{b})$ that translates the analysis result belongs to that region, we say that the analysis is not conclusive. This point reflects only the trend of the situation analyzed. Let us look in detail at one of its sub-regions, as an example.

OFSL sub-region: $0.5 \leq a < 0.8$ and $0.5 \leq b \leq 1$; $0 \leq G < 0.6$ and $-0.5 \leq H < 0$.

In this sub-region, we have a relatively small situation of inconsistency and falsity, but closer to total inconsistency (point B) than to total falsity (point D). Therefore, we say that it is a sub-region of almost inconsistency tending to falsity.

Let us observe, therefore, that the USCP divided into twelve regions allows us to analyze the logical state of a proposition of logic Eτ represented by the point $X = (\boldsymbol{a}; \boldsymbol{b})$. That is why this configuration was called the **para-analyzer algorithm** [7].

11.3 Decision Rule

The regions CPQ (region of truth) and DTU (region of falsity) can be called decision regions. The first favorable decision (viability) and the second unfavorable decision (unfeasibility). This means that if, in the analysis of a project, the result takes us

Table 11.1 Analysis summary of twelve USCP regions

Region	a	b	G	H	Description	Representation
AMN	[0; 0.4]	[0; 0.4]	[−1; −0.6]	[−0.4; 0.4]	Paracompleteness	⊥
BRS	[0.6; 1]	[0.6; 1]	[0.6; 1]	[−0.4; 0.4]	Inconsistency	T
CPQ	[0.6; 1]	[0; 0.4]	[−0.4; 0.4]	[0.6; 1]	Truth	V
DTU	[0; 0.4]	[0.6; 1]	[−0.4; 0.4]	[−1; −0.6]	Falsity	F
OFSL	[0.5; 0.8 [	[0.5; 1]	[0; 0.6 [	[−0.5; 0 [	Quasi inconsistency tending to falsity	QT → F
OHUL	] 0.2; 0.5 [	[0.5; 1]	[0; 0.5 [	] −0.6; 0 [	Quasi falsity tending to inconsistency	QF → T
OHTI	[0; 0.5 [	[0.5; 0.8 [	[−0.5; 0 [	] −0.6; 0 [	Quasi falsity tending to paracompleteness	QF → ⊥
OENI	[0; 0.5 [	] 0.2; 0.5[	] −0.6; 0 [	] −0.5; 0 [	Quasi paracompleteness tending to falsity	Q⊥ → F
OEMK	] 0.2; 0.5 [	[0; 0.5 [	] −0.6; 0 [	[0; 0.5 [	Quasi paracompleteness tending to truth	Q⊥ → V
OGPK	[0.5; 0.8 [	[0; 0.5 [	[−0.5; 0 [	[0; 0.6 [	Quasi truth tending to paracompleteness	QV → ⊥
OGQJ	[0.5; 1]	] 0.2; 0.5 [	[0; 0.5 [	[0; 0.6 [	Quasi truth tending to inconsistency	QV → T
OFRJ	[0.5; 1]	[0.5; 0.8 [	[0; 0.6 [	[0; 0.5]	Quasi inconsistency tending to truth	QT → V

to a point in the CPQ region (truth), the decision is favorable; if, at a point in the DTU region (falsity), the decision is unfavorable to the enterprise; but if the result takes us to a point in any region other than these two, we say that the analysis is not conclusive. We can therefore state the following decision rule:

H ≥ 0.60 ⇒ favorable decision (the project is viable);
H ≤ – 0.60 ⇒ unfavorable decision (the project is not viable);
– 0.60 < H < 0.60 ⇒ not conclusive analysis.

We are adopting |H| = 0.60 as limit lines of truth and falsity. This means that the analysis is only conclusive when |H| ≥ 0.60. Therefore, this value reflects the level of requirement (Lreq) of the analysis. That said, the decision rule can be represented in the following more general way:

H ≥ Lreq ⇒ favorable decision (the project is viable);
H ≤ – Lreq ⇒ unfavorable decision (the project is not viable);
– Lreq < H < Lreq ⇒ non conclusive analysis.

If the result belongs to the BRS region (region of inconsistency), the analysis is not conclusive as to the viability of the enterprise, but it shows a high degree of inconsistency of the data ($G \geq 0.60$). Similarly, if it belongs to the AMN region (paracompleteness), it means that the data present a high degree of paracompleteness or lack of information ($G \leq -0.60$).

Let us note that the level of requirement depends on the security, the confidence that one wants to have in the decision, which, in turn, depends on the responsibility that it implies, the investment that is at stake, the involvement or not of risk to human lives, etc.

11.4 NOT, OR and AND Operators of Logic Eτ

NOT is defined by: NOT $(\boldsymbol{a}; \boldsymbol{b}) = (\boldsymbol{b}; \boldsymbol{a})$. The NOT operator must correspond to the negation of the annotated logic.

Note that: NOT (T) = T, NOT (⊥) = ⊥, NOT (V) = F and NOT (F) = V.

The **OR** and **AND** operators are defined as follows:

$(\boldsymbol{a}_1; \boldsymbol{b}_1)$ OR $(\boldsymbol{a}_2; \boldsymbol{b}_2)$ OR ... OR $(\boldsymbol{a}_n; \boldsymbol{b}_n) = (\max\{\boldsymbol{a}_1, \boldsymbol{a}_2, \ldots, \boldsymbol{a}_n\}; \max\{\boldsymbol{b}_1, \boldsymbol{b}_2, \ldots, \boldsymbol{b}_n\})$.

$(\boldsymbol{a}_1; \boldsymbol{b}_1)$ AND $(\boldsymbol{a}_2; \boldsymbol{b}_2)$ AND ... AND $(\boldsymbol{a}_n; \boldsymbol{b}_n) = (\min\{\boldsymbol{a}_1, \boldsymbol{a}_2, \ldots, \boldsymbol{a}_n\}; \min\{\boldsymbol{b}_1, \boldsymbol{b}_2, \ldots, \boldsymbol{b}_n\})$.

The **OR** operator has the same sense as the classic disjunction, that is, the sense of maximizing the components of the annotations. Therefore, it should be applied in situations in which the two or more items surveyed are not all determinants, as long as one of them has a favorable condition to consider the analysis result satisfactory.

The **AND** operator has the same meaning as the classic conjunction, that is, the meaning of minimizing the components of the annotations. Therefore, it must be applied in situations in which the two or more researched items are all determinants, being essential that all present favorable conditions so that the result of the analysis can be considered satisfactory.

Usually, what is done, when designing the analysis of a real situation, is to separate the items surveyed into groups. These must be assembled in such a way that [8].

(a) the existence of one item from each group with a favorable condition is sufficient to consider the research result as satisfactory;
(b) there are as many groups as there is a minimum number of items that must have favorable conditions to consider the result of the research as satisfactory.

Having made this division, the **OR** operator is applied within each group (intra-groups) and then the **AND** operator between the results obtained in the groups (inter-groups). For example, the application of the **OR** and **AND** operators in the analysis of the opinions of four specialists can be outlined as follows:

[(Specialist 1) **OR** (Specialist 2)] **AND** [(Specialist 3) **OR** (Specialist 4)].

11.5 The Decision Making Process: Paraconsistent Decision-Making Method (PDM)

To present the method, we will apply it to a real problem (case) in which the decision to open the Engineering course in the city of São José dos Campos (SJC) by the Universidade Paulista (UNIP) was studied. This is a decision that university administrators are constantly faced with.

Many factors influence such a decision. They can be legal, social, economic, etc. [3]. Briefly, we will anticipate that, to apply the PDM, we must select the factors (F_i) of greatest influence in the decision and establish the sections (S_j) that translate the conditions in which each one can be found in real situations. Then, we call on specialists (E_k) to assign weights to the factors and degrees of favorable evidence and contrary evidence to the success of the undertaking, for each of the factors chosen, under the conditions translated by the established sections. These values attributed by the specialists will constitute the database.

Then, through applications of the maximization (**OR**) and minimization (**AND**) techniques of logic Eτ, we obtain a favorable degree of favorable evidence ($\boldsymbol{a}_{i,R}$) and a degree of contrary evidence ($\boldsymbol{b}_{i,R}$), resulting, for each factor. Each pair ($\boldsymbol{a}_{i,R}$; $\boldsymbol{b}_{i,R}$) results, in the USCP, in a point that represents the influence of the factor in the decision.

For the final decision making by the administrator, it is not enough to know how each factor influences, but the joint influence of all the chosen factors is important. This can be determined by the center of gravity or barycenter, **W**, of the points that represent them, separately, in the USCP. The degree of favorable evidence ($\boldsymbol{a}_W$) of W is the weighted average of the degrees of favorable evidence resulting ($\boldsymbol{a}_{i,R}$) from all factors, and its degree of contrary evidence ($\boldsymbol{b}_W$), the weighted average of the resulting degrees of contrary evidence ($\boldsymbol{b}_{i,R}$) [7]. With these values we can calculate the degree of certainty of **W** and apply the decision rule.

11.5.1 The Stages of the PDM

To methodize the decision-making process, we considered eight stages for the application of PDM. In this real application that we are going to analyze, opening of the Engineering course in SJC by UNIP, we will follow each stage, one by one.

1st stage: setting the requirement level. The requirement level is fixed according to the case to be analyzed. If, it implies great responsibility (high investments, risk of human lives, etc.), it needs to be set at a high value. In this application, after exchanging ideas with the leaders of UNIP, we decided to set it at 0.50. This means that, in order to make a decision (positive or negative) the module of the difference between the degrees of favorable and unfavorable evidence must be at least equal to 0.50. With the determination of the requirement level, the decision rule is defined, and the para-analyzer algorithm is configured (Fig. 11.3).

(a)

H ≥ 0.5 : Favorable decision (the enterprise is viable)

H ≤ − 0.5: Unfavorable decision (the enterprise is not viable)

-- 0.5 < H < 0.5: Non conclusive analysis

(b)

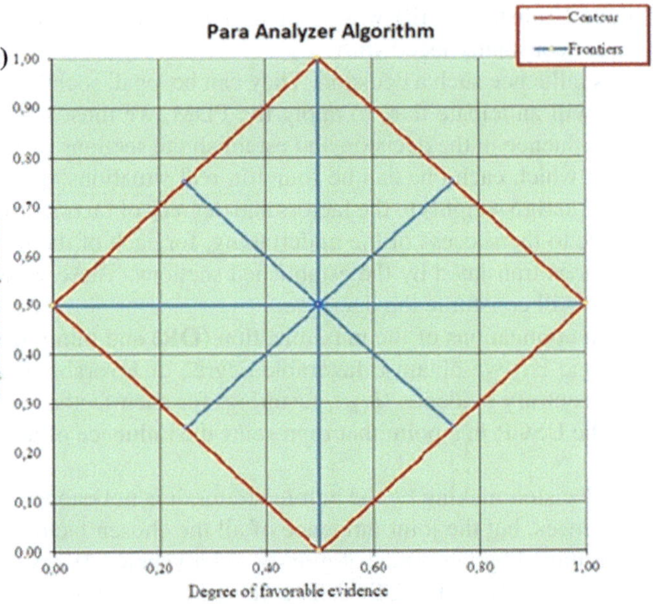

Fig. 11.3 **a** The decision rule. **b** The para analyzer algorithm

2nd stage: choosing the factors (F_i) of greatest influence in the enterprise.

3rd stage: establishment of the sections (S_j) for each of the factors.

Analyzing all the factors (social, legal, economic, etc.) that influence the success (or failure) of a higher education course, we chose twelve factors ($1 \leq i \leq 12$); for each factor, considering the precision and refinement we wanted for the analysis, we established three Sections ($1 \leq j \leq 3$). Section S_1 reflects a condition in which the factor is favorable to the enterprise; S_2, an indifferent condition; and S_3 an unfavorable condition. Thus, the factors chosen, and the sections established are as follows.

F01: applicant / vacancy ratio (C / V) of course X in the selection exams of region Y under study.

S1: C / V > 4; S2: 2 ≤ C / V ≤ 4; S3: C / V < 2.

F02: number of high school graduates (Nc) in region Y.

S1: Nc > 2 V; S2: V ≤ Nc ≤ 2 V; S3: Nc < V,

V = number of places offered for higher education in region Y.

F03: number of jobs (Nj) offered annually in region Y.

S1: Nj > 2F; S2: F ≤ Nj ≤ 2F;S3: Nj < F.

F = annual number of trainees in higher education in region Y.

F04: average monthly family income (Rf) of the population in region Y.
S1: Rf > US \$ 6,000; S2: \$ 2,000 $\leq$ Rf $\leq$ \$ 6,000; S3: Rf < \$ 2,000.
F05: average annual index (Ia) of withdrawals from the course.
S1: Ia < 10%; S2: 10% $\leq$ Ia $\leq$ 40%; S3: Ia > 40%.
F06: demographic density (DD) of the region.
S1: high: DD > 400 inhabitants / km^2; S3: low: DD < 100 inhabitants / km^2.
S2: average: 100 inhabitants / km^2 $\leq$ DD $\leq$ 400 inhabitants / km^2;
F07: cost of investments in fixed assets (C).
S1: C < 75% Ra; S2: 75% Ra $\leq$ C $\leq$ 125% Ra; S3: C > 125% Ra.
Ra = expected annual revenue for course X.
F08: concept of the institution with the community.
S1: concept A or B; S2: concept C; S3: concept D or E.
F09: monthly cost of teachers (Cmp).
S1: Cmp < 25% Rm; S2: 25% Rm $\leq$ Cmp $\leq$ 45% Rm; S3: Cmp > 45% Rm.
Rm = estimated monthly revenue for course X.
F10: monthly tuition (Mc) for course X.
S1: Mc < 80% Mm; S2: 80% Mm $\leq$ Mc $\leq$ 120% Mm; S3: Mc > 120% Mm.
Mm = average tuition for course X (or similar) in other schools in region Y.
F11: average number of students per class (Nas).
S1: Nas > 80; S2: 50 $\leq$ Nas $\leq$ 80; S3: Nas < 50.
F12: average number of employees per class (Nec).
S1: Nec < 5; S2: 5 $\leq$ Nec $\leq$ 10; S3: Nec > 10.

4th stage: construction of the database. In this work, we will admit all factors with the same weight, equal to 1 (one). We chose eight specialists with different backgrounds, but all linked to higher education and with great experience in the pedagogical and administrative areas of private schools. All are PhDs in Education, Engineering, Physics or Mathematics; all with experience in at least two of the following activities: course coordination, department head, institute and pro-rectory or university rectory. They are: E1: Ph.D. in Education and University Dean; E2: Ph.D. in Engineering and University Dean; E3: Ph.D. in Engineering and Director of the exact area at university; E4: Ph.D. in Engineering and coordinator of a university course; E5: PhD in Physics and Engineering and Director of the exact area at university; E6: Ph.D. in Engineering, department head and course coordinator; E7: Master of Engineering and Engineering College Director; E8: Ph.D. in Mathematics and Director of the Exact Sciences Area at university.

To the specialists, we present a form with the chosen influencing factors and the established sections, also containing explanations about the method and guidelines on how to fill it out. Based on the experts' responses, that is, the values of the degrees of favorable and contrary evidence that they attributed to the factors in the established sections, we built the database in Table 11.2.

5th stage: field research. This research consists of verifying, by collecting real data and information, in which section each factor is found. Here, we will omit the details of this information and present the result only, which is summarized in Table 11.3.

Table 11.2 Database

Factor	Section	E1		E2		E3		E4		E6		E5		E8		E7	
		$a_{i,1}$	$b_{i,1}$	$a_{i,2}$	$b_{i,2}$	$a_{i,3}$	$b_{i,3}$	$a_{i,4}$	$b_{i,4}$	$a_{i,6}$	$b_{i,6}$	$a_{i,5}$	$b_{i,5}$	$a_{i,8}$	$b_{i,8}$	$a_{i,7}$	$b_{i,7}$
F_{01}	S_1	0.90	0.08	0.60	0.40	0.90	0.10	0.98	0.05	0.90	0.15	0.95	0.50	0.90	0.10	0.94	0.23
	S_2	0.40	0.40	0.60	0.60	0.60	0.40	0.65	0.28	0.50	0.40	0.45	0.50	0.50	0.50	0.56	0.47
	S_3	0.07	0.90	0.30	0.70	0.40	0.40	0.40	0.63	0.30	0.95	0.15	0.80	0.20	0.50	0.14	0.95
F_{02}	S_1	0.80	0.15	0.60	0.40	0.80	0.10	0.93	0.05	0.95	0.20	0.75	0.40	0.95	0.10	0.76	0.25
	S_2	0.30	0.20	0.50	0.50	0.60	0.30	0.55	0.48	0.35	0.60	0.30	0.80	0.75	0.25	0.48	0.44
	S_3	0.10	0.85	0.30	0.70	0.40	0.60	0.38	0.77	0.25	0.88	0.10	0.90	0.30	0.70	0.11	0.91
F_{03}	S_1	0.80	0.15	0.70	0.30	0.95	0.10	0.95	0.08	0.92	0.12	0.90	0.20	0.90	0.10	0.89	0.17
	S_2	0.30	0.65	0.50	0.50	0.80	0.15	0.60	0.36	0.38	0.50	0.40	0.70	0.75	0.25	0.43	0.42
	S_3	0.10	0.90	0.30	0.70	0.60	0.30	0.25	0.87	0.20	0.94	0.15	0.90	0.50	0.60	0.13	0.94
F_{04}	S_1	0.95	0.10	0.80	0.20	0.90	0.15	0.85	0.25	0.98	0.10	0.99	0.10	0.90	0.20	0.98	0.13
	S_2	0.40	0.25	0.60	0.40	0.60	0.20	0.52	0.50	0.45	0.70	0.45	0.50	0.50	0.50	0.60	0.34
	S_3	0.10	0.75	0.30	0.70	0.40	0.50	0.18	0.85	0.18	0.97	0.15	0.90	0.30	0.80	0.23	0.68
F_{05}	S_1	0.80	0.15	0.70	0.30	0.80	0.30	0.80	0.35	0.96	0.14	0.97	0.05	0.95	0.00	0.92	0.13
	S_2	0.50	0.50	0.40	0.60	0.60	0.30	0.50	0.50	0.33	0.60	0.50	0.60	0.70	0.60	0.45	0.61
	S_3	0.10	0.85	0.20	0.80	0.40	0.40	0.30	0.73	0.26	0.89	0.20	0.80	0.00	0.95	0.14	0.97
F_{06}	S_1	0.90	0.10	0.60	0.40	0.80	0.10	0.85	0.10	0.91	0.20	0.95	0.15	0.80	0.20	0.92	0.20
	S_2	0.50	0.50	0.60	0.40	0.70	0.20	0.60	0.45	0.29	0.60	0.60	0.45	0.50	0.50	0.55	0.49
	S_3	0.10	0.90	0.40	0.60	0.60	0.30	0.45	0.48	0.31	0.92	0.20	0.90	0.40	0.60	0.23	0.96

(continued)

Table 11.2 (continued)

Factor	Section	E1		E2		E3		E4		E6		E5		E8		E7	
		$a_{i,1}$	$b_{i,1}$	$a_{i,2}$	$b_{i,2}$	$a_{i,3}$	$b_{i,3}$	$a_{i,4}$	$b_{i,4}$	$a_{i,6}$	$b_{i,6}$	$a_{i,5}$	$b_{i,5}$	$a_{i,8}$	$b_{i,8}$	$a_{i,7}$	$b_{i,7}$
F_{07}	S_1	0.85	0.10	0.70	0.30	0.90	0.20	0.85	0.30	0.96	0.19	0.85	0.20	0.70	0.30	0.85	0.32
	S_2	0.40	0.40	0.60	0.40	0.70	0.25	0.55	0.40	0.37	0.51	0.35	0.80	0.30	0.70	0.39	0.70
	S_3	0.10	0.87	0.30	0.70	0.50	0.50	0.40	0.55	0.29	0.92	0.15	0.95	0.10	0.90	0.11	0.98
F_{08}	S_1	0.90	0.15	0.60	0.40	0.95	0.05	0.75	0.35	0.94	0.11	0.99	0.10	0.70	0.10	0.94	0.13
	S_2	0.50	0.50	0.50	0.50	0.70	0.20	0.50	0.50	0.44	0.58	0.50	0.50	0.50	0.30	0.43	0.49
	S_3	0.07	0.95	0.30	0.70	0.50	0.50	0.40	0.55	0.18	0.92	0.10	0.90	0.20	0.80	0.18	0.97
F_{09}	S_1	0.80	0.15	0.70	0.30	0.80	0.15	0.60	0.50	0.97	0.16	0.95	0.20	0.70	0.30	0.91	0.22
	S_2	0.40	0.35	0.60	0.40	0.50	0.50	0.55	0.40	0.31	0.50	0.50	0.50	0.40	0.60	0.39	0.62
	S_3	0.10	0.90	0.40	0.60	0.10	0.80	0.45	0.37	0.28	0.93	0.20	0.90	0.20	0.80	0.05	0.99
F_{10}	S_1	0.90	0.15	0.80	0.40	0.90	0.20	0.98	0.01	0.99	0.11	0.98	0.10	0.70	0.20	0.99	0.13
	S_2	0.45	0.40	0.60	0.40	0.70	0.25	0.75	0.30	0.39	0.57	0.45	0.65	0.50	0.50	0.26	0.65
	S_3	0.10	0.85	0.60	0.40	0.60	0.40	0.20	0.90	0.18	0.94	0.25	0.85	0.50	0.70	0.03	0.93
F_{11}	S_1	0.90	0.15	0.60	0.40	0.80	0.15	0.70	0.40	0.96	0.14	0.10	0.95	0.80	0.20	0.90	0.19
	S_2	0.50	0.30	0.70	0.30	0.70	0.25	0.55	0.45	0.43	0.59	0.40	0.60	0.60	0.30	0.55	0.44
	S_3	0.15	0.85	0.30	0.70	0.60	0.35	0.40	0.50	0.29	0.93	0.90	0.15	0.50	0.60	0.12	0.91
F_{12}	S_1	0.90	0.07	0.60	0.40	0.80	0.15	0.70	0.40	0.91	0.40	0.90	0.10	0.70	0.10	0.92	0.17
	S_2	0.50	0.50	0.60	0.40	0.70	0.25	0.55	0.45	0.42	0.60	0.30	0.75	0.40	0.60	0.58	0.50
	S_3	0.15	0.90	0.30	0.70	0.60	0.35	0.40	0.50	0.24	0.90	0.15	0.90	0.20	0.80	0.12	0.90

Table 11.3 Search results for an Engineering course in the SJC region

Factor	F_i	F_1	F_2	F_3	F_4	F_5	F_6	F_7	F_8	F_9	F_{10}	F_{11}	F_{12}
Section	S_j	S_1	S_1	S_1	S_2	S_1	S_1	S_1	S_2	S_1	S_3	S_2	S_1

6th stage: Obtain the degrees of favorable evidence ($a_{i,R}$) and contrary evidence ($b_{i,R}$), resulting for the factors, under the conditions of the sections obtained in the research. For this, we must distribute the specialists into groups, in the most appropriate way. Considering their backgrounds and experiences, we distribute them into the following groups. Group A: E1; group B: E2 and E3; group C: E4 and E6; group D: E5, E7 and E8. With this formation of groups, the application of the maximization (**OR** operator) and minimization (**AND** operator) rules is outlined as follows:

[E1] AND [E2 OR E3] AND [E4 OR E6] AND [E5 OR E7 OR E8] or.

[GA] and [GB] and [GC] and [GD]

That done, we should take the search result to column 2 of the PDM calculation table (Table 11.4, which, for ease of reading, was divided into two parts: Tables 11.6 and 11.7).

Then, the PDM (i) searches for the corresponding degrees of favorable and contrary evidence in the database (Table 11.1) and places them in columns 3 to 18 of Tables 11.4 or 11.6; (ii) apply the maximization rule within groups (intra-groups), obtaining columns 19 to 26 in Tables 11.5 or 11.7; (iii) applies the minimization rule between the results obtained in each group (inter groups), obtaining columns 27 and 28, with the degrees of favorable evidence ($\boldsymbol{a}_{i,R}$) and contrary evidence ($\boldsymbol{b}_{i,R}$), resulting for the factors, in the conditions of the sections obtained in the research.

With these values, the PDM (iv) calculates the degrees of certainty and contradiction of the factors (columns 29 and 30) and (v) applies the decision rule, comparing the degrees of certainty with the level of requirement, and informs if, according to each factor, the project is viable, unfeasible or if the factor is not conclusive (column 31 of Tables 11.5 or 11.7).

7th stage: Obtain the degree of favorable evidence (a_W) and the degree of contrary evidence (b_W) of the barycenter. The PDM (vi) calculates the average (in this case, arithmetic, since we are considering that the factors have equal weights) of the degrees of favorable evidence ($\boldsymbol{a}_{i,R}$) and the degrees of contrary evidence ($\boldsymbol{b}_{i,R}$), resulting for the factors, obtaining the corresponding values for the barycenter (last row of columns 27 and 28 of Tables 11.5 or 11.7). Thus, we obtained $\boldsymbol{a}_W = 0.70$ and $\boldsymbol{b}_W = 0.18$. With these values, the program calculates the degree of certainty of the barycenter (last row in column 29): $H = \boldsymbol{a}_W - \boldsymbol{b}_W = 0.70 - 0.18 = 0.52$.

8th stage: decision making. Once the degree of certainty of the barycenter has been calculated, the PDM applies the decision rule, comparing it with the level of requirement adopted. As $0.52 \geq 0.50$, we concluded that the project is viable at the 0.50 requirement level (last row in column 31). In other words, the opening of the Engineering course at SJC is a viable enterprise at the 50% requirement level.

Table 11.4 PDM calculation table (1st part)

Factor (column 1)		Section (2)	Grup A		Grup B				Grup C				Grup D					
			E1 (3 and 4)		E2 (5 and 6)		E3 (7 and 8)		E4 (9 and 10)		E6 (11 and 12)		E5 (13 and 14)		E8 (15 and 16)		E7 (17 and 18)	
Fi	***Pi***	***Spi***	***ai,1***	***bi,1***	***ai,2***	***bi,2***	***ai,3***	***bi,3***	***ai,4***	***bi,4***	***ai,5***	***bi,5***	***ai,6***	***bi,6***	***ai,6***	***bi,6***	***ai,7***	***bi,7***
F01	1	S1	0.87	0.08	0.60	0.40	0.89	0.10	0.98	0.05	0.94	0.50	0.90	0.15	0.94	0.23	0.87	0.10
F02	1	S1	0.76	0.15	0.61	0.40	0.82	0.10	0.93	0.05	0.75	0.40	0.95	0.20	0.76	0.25	0.95	0.10
F03	1	S1	0.78	0.15	0.74	0.30	0.95	0.10	0.95	0.08	0.90	0.20	0.92	0.12	0.89	0.17	0.93	0.10
F04	1	S2	0.70	0.10	0.60	0.40	0.70	0.20	0.52	0.50	0.70	0.50	0.70	0.70	0.60	0.34	0.50	0.50
F05	1	S1	0.80	0.15	0.73	0.30	0.81	0.30	0.80	0.35	0.97	0.05	0.96	0.14	0.92	0.13	0.95	0.00
F06	1	S1	0.89	0.10	0.62	0.40	0.80	0.10	0.85	0.10	0.95	0.15	0.91	0.20	0.92	0.20	0.80	0.20
F07	1	S1	0.85	0.10	0.72	0.30	0.90	0.20	0.85	0.30	0.85	0.20	0.96	0.19	0.85	0.32	0.72	0.30
F08	1	S2	0.60	0.50	0.50	0.50	0.70	0.20	0.50	0.50	0.60	0.50	0.44	0.58	0.43	0.49	0.60	0.30
F09	1	S1	0.82	0.15	0.71	0.30	0.79	0.15	0.60	0.50	0.96	0.20	0.97	0.16	0.91	0.22	0.70	0.30
F10	1	S3	0.10	0.85	0.60	0.40	0.60	0.40	0.20	0.90	0.25	0.85	0.18	0.94	0.03	0.93	0.50	0.70
F11	1	S2	0.60	0.30	0.70	0.30	0.70	0.25	0.60	0.45	0.40	0.60	0.43	0.59	0.55	0.44	0.60	0.30
F12	1	S1	0.95	0.07	0.65	0.40	0.77	0.15	0.70	0.40	0.90	0.10	0.91	0.40	0.92	0.17	0.68	0.10
	12		Baricenter W: averages of the resultant degrees															

Table 11.5 PDM calculation table (2nd part)

Grup A (19 and 20)		Grup B (21 and 22)		Grup C (23 and 24)		Grup D (25 and 26)		A AND B AND C AND D (27 and 28)		Level of requirement =		0.50
E1		E2 OR E3		E4 OR E6		E5 OR E8 OR E7				Conclusions (29, 30 and 31)		
ai,1	***bi,1***	***ai,gB***	***bi,gB***	***ai,gC***	***bi,gC***	***ai,gD***	***bi,gD***	***ai,R***	***bi,R***	**H**	**G**	**Decision**
0.87	0.08	0.89	0.40	0.98	0.50	0.94	0.23	0.87	0.08	0.79	−0.05	VIABLE
0.76	0.15	0.82	0.40	0.93	0.40	0.95	0.20	0.76	0.15	0.61	−0.09	VIABLE
0.78	0.15	0.95	0.30	0.95	0.20	0.93	0.12	0.78	0.12	0.66	−0.10	VIABLE
0.70	0.10	0.70	0.40	0.70	0.50	0.70	0.70	0.70	0.10	0.60	−0.20	VIABLE
0.80	0.15	0.81	0.30	0.97	0.35	0.96	0.14	0.80	0.14	0.66	−0.06	VIABLE
0.89	0.10	0.80	0.40	0.95	0.15	0.91	0.20	0.80	0.10	0.70	−0.10	VIABLE
0.85	0.10	0.90	0.30	0.85	0.30	0.96	0.30	0.85	0.10	0.75	−0.05	VIABLE
0.60	0.50	0.70	0.50	0.60	0.50	0.60	0.58	0.60	0.50	0.10	0.10	NOT CONCLUSIVE
0.82	0.15	0.79	0.30	0.96	0.50	0.97	0.30	0.79	0.15	0.64	−0.06	VIABLE
0.10	0.85	0.60	0.40	0.25	0.90	0.50	0.94	0.10	0.40	−0.30	−0.50	NOT CONCLUSIVE
0.60	0.30	0.70	0.30	0.60	0.60	0.60	0.59	0.60	0.30	0.30	−0.10	NOT CONCLUSIVE
0.95	0.07	0.77	0.40	0.90	0.40	0.91	0.40	0.77	0.07	0.70	−0.16	VIABLE
Baricenter W: averages of the resultant degrees								0.70	0.18	0.52	−0.11	VIABLE

Table 11.6 PDM calculation table (1st part)

Factor (column 1)		Section (2)	Grup A		Grup B				Grup C				Grup D					
			E1 (3 and 4)		E2 (5 and 6)		E3 (7 and 8)		E4 (9 and 10)		E6 (11 and 12)		E5 (13 and 14)		E8 (15 and 16)		E7 (17 and 18)	
Fi	***Pi***	***Spi***	***ai,1***	***bi,1***	***ai,2***	***bi,2***	***ai,3***	***bi,3***	***ai,4***	***bi,4***	***ai,5***	***bi,5***	***ai,6***	***bi,6***	***ai,6***	***bi,6***	***ai,7***	***bi,7***
F01	1	S1	0.87	0.08	0.60	0.40	0.89	0.10	0.98	0.05	0.94	0.50	0.90	0.15	0.94	0.23	0.87	0.10
F02	1	S1	0.76	0.15	0.61	0.40	0.82	0.10	0.93	0.05	0.75	0.40	0.95	0.20	0.76	0.25	0.95	0.10
F03	1	S1	0.78	0.15	0.74	0.30	0.95	0.10	0.95	0.08	0.90	0.20	0.92	0.12	0.89	0.17	0.93	0.10
F04	1	S2	0.70	0.10	0.60	0.40	0.70	0.20	0.52	0.50	0.70	0.50	0.70	0.70	0.60	0.34	0.50	0.50
F05	1	S1	0.80	0.15	0.73	0.30	0.81	0.30	0.80	0.35	0.97	0.05	0.96	0.14	0.92	0.13	0.95	0.00
F06	1	S1	0.89	0.10	0.62	0.40	0.80	0.10	0.85	0.10	0.95	0.15	0.91	0.20	0.92	0.20	0.80	0.20
F07	1	S1	0.85	0.10	0.72	0.30	0.90	0.20	0.85	0.30	0.85	0.20	0.96	0.19	0.85	0.32	0.72	0.30
F08	1	S2	0.60	0.50	0.50	0.50	0.70	0.20	0.50	0.50	0.60	0.50	0.44	0.58	0.43	0.49	0.60	0.30
F09	1	S1	0.82	0.15	0.71	0.30	0.79	0.15	0.60	0.50	0.96	0.20	0.97	0.16	0.91	0.22	0.70	0.30
F10	1	S3	0.10	0.85	0.60	0.40	0.60	0.40	0.20	0.90	0.25	0.85	0.18	0.94	0.03	0.93	0.50	0.70
F11	1	S2	0.60	0.30	0.70	0.30	0.70	0.25	0.60	0.45	0.40	0.60	0.43	0.59	0.55	0.44	0.60	0.30
F12	1	S1	0.95	0.07	0.65	0.40	0.77	0.15	0.70	0.40	0.90	0.10	0.91	0.40	0.92	0.17	0.68	0.10
	12		**Baricenter W: averages of the resultant degrees**															

Table 11.7 PDM calculation table (2nd part)

Grup A (19 and 20)		Grup B (21 and 22)		Grup C (23 and 24)		Grup D (25 and 26)		A AND B AND C AND D (27 and 28)		Level of requirement		0.50
E1		E2 OR E3		E4 OR E6		E5 OR E8 OR E7				Conclusions (29, 30 and 31)		
ai,1	***bi,1***	***ai,gB***	***bi,gB***	***ai,gC***	***bi,gC***	***ai,gD***	***bi,gD***	***ai,R***	***bi,R***	**H**	**G**	**Decision**
0.87	0.08	0.89	0.40	0.98	0.50	0.94	0.23	0.87	0.08	0.79	−0.05	VIABLE
0.76	0.15	0.82	0.40	0.93	0.40	0.95	0.20	0.76	0.15	0.61	−0.09	VIABLE
0.78	0.15	0.95	0.30	0.95	0.20	0.93	0.12	0.78	0.12	0.66	−0.10	VIABLE
0.70	0.10	0.70	0.40	0.70	0.50	0.70	0.70	0.70	0.10	0.60	−0.20	VIABLE
0.80	0.15	0.81	0.30	0.97	0.35	0.96	0.14	0.80	0.14	0.66	−0.06	VIABLE
0.89	0.10	0.80	0.40	0.95	0.15	0.91	0.20	0.80	0.10	0.70	−0.10	VIABLE
0.85	0.10	0.90	0.30	0.85	0.30	0.96	0.30	0.85	0.10	0.75	−0.05	VIABLE
0.60	0.50	0.70	0.50	0.60	0.50	0.60	0.58	0.60	0.50	0.10	0.10	NOT CONCLUSIVE
0.82	0.15	0.79	0.30	0.96	0.50	0.97	0.30	0.79	0.15	0.64	−0.06	VIABLE
0.10	0.85	0.60	0.40	0.25	0.90	0.50	0.94	0.10	0.40	−0.30	−0.50	NOT CONCLUSIVE
0.60	0.30	0.70	0.30	0.60	0.60	0.60	0.59	0.60	0.30	0.30	−0.10	NOT CONCLUSIVE
0.95	0.07	0.77	0.40	0.90	0.40	0.91	0.40	0.77	0.07	0.70	−0.16	VIABLE
Baricenter W: averages of the resultant degrees								**0.70**	**0.18**	**0.52**	**−0.11**	VIABLE

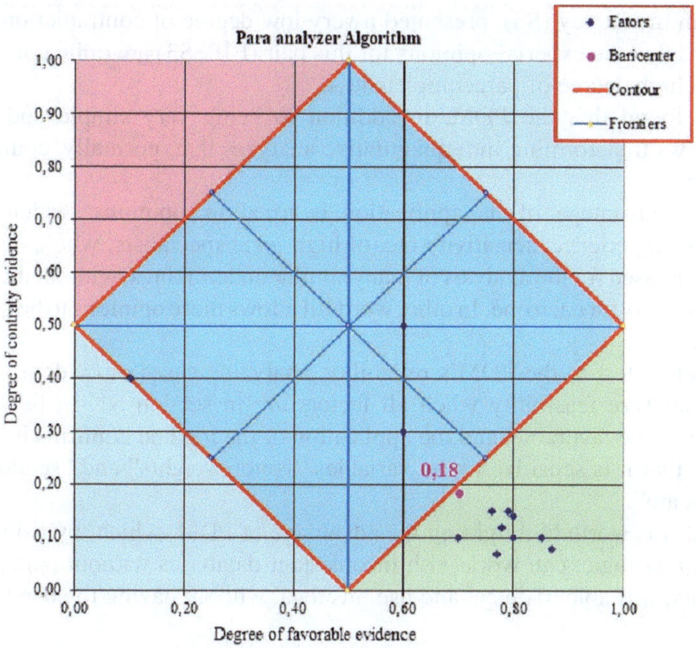

Fig. 11.4 Analysis of the result by the para-analyzer algorithm

11.5.2 Analysis of Results

The analysis made with the aid of the PDM calculation table, can also be observed by the para-analyzer algorithm (Fig. 11.4).

11.6 Conclusions and Observations

Looking at Tables 11.5 or 11.7, we note that, except for factors F08, F10 and F11, which were not conclusive, the nine other factors influenced the indication of the viability of the project (favorable decision), at the 0.50 requirement level.

Two observed facts caught our attention. The first is that the F04 factor, even being in the condition in which it should be indifferent (section S2), accused the viability of opening the course. Probably, the experts did not agree with the decision-making coordinator regarding what he considered an indifferent condition for this factor, that is, what the coordinator thought was an indifferent condition the experts did not find.

The second fact that drew attention was that the F10 factor, even though it was in the condition defined by section S3 (unfavorable condition), did not accuse the undertaking of its unfeasibility. In this case, considerations analogous to those made in relation to the first fact are applicable. In addition, the F10 factor, in the section obtained in the survey (S3), presented a very low degree of contradiction (– 0.50). This shows that the experts' opinions for this pair (F10, S3) are quite contradictory, showing a high degree of paracompleteness.

It was found that the PDM, in addition to being very simple and quick to apply, allows transforming into quantitative analyzes that, normally, could only be qualitative.

Another advantage of its application is to allow opinions, which translate knowledge, experience, sensitivity etc. of high-level specialists, who constitute the database, are used for the analysis of many similar undertakings, without them having to expend efforts for each one. In other words, it allows these opinions to be practically perpetuated.

We could attest to the PDM's reliability, analyzing cases of evident result. For example, analyze feasibility when all factors are in section S1. In this case, the decision must be favorable, and the application of the method confirms it. We could also show that it is sensitive to the variables "factor weights" and "sections where the factors are".

Finally, it is worth highlighting the advantage of PDM, which, by using tools of paraconsistent logic, can work with inconsistent databases without collapsing. On the contrary, it is able to detect and use this data, without having to abandon it.

References

1. Abe, Jair Minoro.: Paraconsistent Intelligent Based-Systems: New Trends in the Applications of Paraconsistency, editor, Book Series: "Intelligent Systems Reference Library", Springer-Verlag, Vol. 94, p. 306, ISBN:978-3-319-19721-0, Germany (2015). https://doi.org/10.1007/978-3-319-19722-7
2. Akama, Seiki.: Towards paraconsistent engineering, intelligent systems reference library, Vol. 110, p. 234 (2016). ISBN: 978-3-319-40417-2 (Print) 978-3-319-40418-9 (Online), Series ISSN 1868-4394, Publisher Springer International Publishing. https://doi.org/10.1007/978-3-319-40418-9
3. BRAZIL.: Ministry of Education and Sport (MEC). Ordinance No. 181, of February 23, 1996. Establishes new authorization procedures for the operation of higher undergraduate courses. Diário Oficial da União (DOU), Brasília, DF, edition of 27 feb. (1996)
4. Da Costa, Newton C.A., Vago, Carlo, Subrahmanian, V.S.: The Paraconsistent Logics Pτ. In: Zeitschrift für Math. Logik und Grundlagen der Math, Bd. 37, pp. 139–148 (1991)
5. Da Costa, Newton C.A., Abe, Jair M., Subrahmanian, V.S.: Remarks on Annotated Logic. In: Zeitschrift für Math. Logik und Grundlagen der Math, Bd. 37, pp. 561–570 (1991)
6. Da Costa, Newton C.A., Abe, Jair M.: Some Recent Applications of Paraconsistent Systems in Artificial Intelligence and Robotics. São Paulo, SP, Brazil: Institute of Advanced Studies (IEA), University of São Paulo (USP), p. 12 (1999)
7. De Carvalho, Fábio R., Abe, Jair M.: A paraconsistent decision-making method. Smart Innovation, Systems and Technologies 87. Springer International Publishing, p. 212 (2018). ISBN 978-3-319-74109-3; ISBN eBook 978-3-319-74110-9; https://doi.org/10.1007/978-3-319-74110-9. Switzerland
8. De Carvalho, Fábio R., Abe, Jair M.: Tomadas de Decisão com Ferramentas da Lógica Paraconsistente Anotada. Editora Blucher, São Paulo, Brazil, p. 220 (2011)

9. Dias Junior, O.P.: Deciding on the basis of inaccurate information. Administration Research Book, vol. 8, no. 4, pp. 69–75 (2001)
10. Nakamatsu, Kazumi, Abe, Jair M., Akama, Seiki.: Introduction to annotated logics - foundations for paracomplete and paraconsistent reasoning, series title intelligent systems reference library, Vol. 88, Publisher Springer International Publishing, Copyright Holder Springer International Publishing Switzerland, 1st edn, p. 190 (2015) eBook ISBN 978-3-319-17912-4, https://doi.org/10.1007/978-3-319-17912-4, Hardcover ISBN 978-3-319-17911-7, Series ISSN 1868-4394

Chapter 12
Annotated Logics and Application—An Overview

Jair Minoro Abe

Contents

Abstract The author comments on the annotated logic, which was born in the late 80 s of the last century and was the main subject of investigation of his career. Abe is considered one of the leading pioneers in the applications of paraconsistent logic. The chapter closes with theoretical and application aspects.

Keywords Annotated logic · Paraconsistent logic · Paracomplete logic · Non-alethic logic · AI

J. M. Abe (✉)
Graduate Program in Production Engineering, Paulista University, São Paulo, Brazil
e-mail: jairabe@uol.com.br

J. M. Abe (ed.), *Advances in Applied Logics*, Intelligent Systems Reference Library 243,
https://doi.org/10.1007/978-3-031-35759-6_12

12.1 Introduction

I met Prof. Akama at the First World Congress on Paraconsistency held in 1997 in Ghent, Belgium. It was a pleasant surprise. I was not aware of his interest in paraconsistent logic. The following year I met Akama again at the Stanislaw Jaskowski Memorial Symposium on Paraconsistent Logic, Logical Philosophy, Mathematics, Informatics, in Torun, Poland, organised by J. Perzanowski.

He told me he knew from my studies on annotated logic through R. Sylvan, who had invited him to spend a season in Japan. Months earlier, Sylvan had just spent time as a visiting professor at the Institute for Advanced Studies, University of São Paulo. I needed details with my doctoral thesis on logic, which activated the challenge of discussing my PhD thesis with him, supported by his great experience.

To my surprise, he became interested in the subject and told me that the issue was so promising that he would investigate it to seek a unification of the logic. Sylvan even asked me to look for a suitable name for this broader and broader logic. Numerous correspondences about a generalisation of the annotated logic were made. A draft was published in [13]. Master Sylvan died unpredictably shortly afterwards.

Akama's contributions vary: they range from the philosophical analysis of non-classical to technical contributions in logical systems, mainly in non-classical logic and AI applications. In addition to the annotated logic, he has contributed to Rough Set Theory with the T. Murai and Y. Kudo group.

It should be observed that one of his favourite activities is to write books for scientific dissemination, in addition to books that include research contributions. He is the author of more than a hundred books, mainly in the area of Informatics. Many bookstores even have a corner dedicated to their readers. In his spare time, he plays musical instruments.

12.1.1 Paraconsistent Logic

Let T be a theory founded on a logic L, and suppose that the language of L and T contains a symbol for the negation (if there is more than one negation, one of them must be chosen due to its logical-formal characteristics). T is said to be inconsistent if it has contradictory theorems, i.e., such that one is the negation of the other; otherwise, T is said to be consistent. T is considered trivial if all formulas of L (or all closed formulas of L) are T theorems; otherwise, T is called non-trivial.

Consistency plays a vital role in classical logic and many categories of logic. Indeed, in most usual logical systems, if a theory T is trivial, then T is inconsistent and vice versa.

A logic L is called Paraconsistent if it can serve as a basis for inconsistent but non-trivial theories.

Another significant concept for what follows is that of Paracomplete logic. A logic L is said to be Paracomplete if it can be the underlying logic of theories in which the law of excluded middle is not valid in the following way. Of two contradictory propositions, one of them is true. A logic is called Paracomplete if there are non-trivial maximal systems to which a given formula and its negation do not belong.

In the literature, there are several Paraconsistent logic systems—relevant logic, discussion logic, logics of inherent ambiguity, and the now classic Cn systems by da Costa. In the late 1980s, Annotated logic was discovered. Annotated logic is a type of Paraconsistent logic. That is, they are "naturally" paraconsistent.

Regarding applications, traditional logic has several applications that can be loosely classified into two categories: (1) Direct applications; and (2) Indirect applications.

Direct applications, the most common, gave rise to logic. Here logic is conceived as a mechanism of inference: in ordinary life and science, we make inferences, models, etc. that is, logic is the study of the structure of propositions and their use in encoding deductive inference.

Indirect applications have a technological nature: applications in Artificial Intelligence, electrical circuits, software engineering, programming, automation and robotics or molecular biology and genetics.

Until the late 1980s, only direct applications of paraconsistent systems were known. Nevertheless, in the late 1980s, the first indirect applications of paraconsistent systems began to appear. We now describe a small amount of its history.

12.1.2 Initial Indirect Applications of Paconsistent Logics

Many applications of Paraconsistent logic in Computer Science are related to situations where inconsistencies arise naturally. They often occur in databases, logic programs, and other formalisms representing data, knowledge, and beliefs. Below, we will give examples showing the usefulness of one of these Paraconsistent systems: annotated logic. It should be noted that the case of paraconsistent programming based on annotated logic constitutes a generalisation of the usual logic programming.

The main initial applications were concerned with the following topics [5]:

1. Inferences about inconsistent knowledge base.
2. Inheritance nets.
3. Object-oriented database.
4. Inferences about an inconsistent knowledge base

Expert systems and systems based on knowledge about a D domain are usually built by programmers who generally know little about D. Programmers operate by consulting a group of domain experts. Thus, if we want to build an expert system of medicine that concerns acid–base disorders, we can consult various physicians and get them to articulate the 'rules of thumb' they use in diagnosing patients.

Having done this, we assume that the doctors' practical norms and/or facts can be conveniently expressed in some logical system.

Unfortunately, experts in a given field of study are often prone to disagreement due to human limitations. For example, given the same observable symptoms, physician M1 might believe that the patient has, in all likelihood, a virus infection. On the other hand, the M2 physician may conclude that the patient has an allergic reaction. Doctor M3, very conservative, can only say that the patient is either suffering from a viral infection or an allergy, but not both. We will inevitably face inconsistency if we use doctors' opinions of M1, M2 and M3 to build our knowledge base.

The critical point is that this inconsistency is natural. Scientists often get into disagreements, and often for fully justified reasons. Furthermore, this is true of almost every profession. Sometimes it is just as important to report that scientists have conflicting opinions about a particular problem or phenomenon as it is to report that the phenomenon is guaranteed to occur. Equally, it is relevant that the most widespread way of building knowledge data i. e., consulting experts in the field of interest) is seriously subject to inconsistencies.

We need to verify that the existence of inconsistencies will not harm those who use the knowledge data.

Classical logic is not suitable for dealing directly with inconsistencies. The non-adequacy in these cases arises because, as is known, in classical logic, if T is an axiomatisation of an inconsistent theory, then every formula F of the language underlying such a theory is a logical consequence of T.

Thus, if BCMI is a piece of the helpful basis of knowledge in consistent medicine and some mischievous individuals introduce two new facts, p and p* (the negation of p), into BCMI to form BCMII, then BCMII would be useless, even though (intuitively speaking) the inconsistency in BCMII has nothing to do with it. Do with the information in BCMI.

12.1.3 Inheritance Nets

Inheritance nets constitute a basic diagrammatic form of knowledge representation. An inheritance network can be seen as an ordered graph whose vertices are classified as entities of two kinds: objects and classes. So we can have a network containing an arc from "Clyde" to "Elephant". Here, Clyde is the name of an object, while Elephant is the name of a class. There can also be arcs from classes to classes (e.g. Elephants are Mammals that can be represented similarly).

Similarly, we can have "negative arcs", an arc that says Bill is not an Elephant or an arc that says Elephants are not Birds, etc. As positive or negative arcs are both present, inheritance nets can contain inconsistencies. Inheritance reasoning deals with the problem of deciding what conclusions can be drawn from a given graph. The main distinctions between reasoning by inheritance and annotated logics are as follows:

(1) reasoning by inheritance does not describe inconsistencies even though such inconsistencies may indicate a difference of opinion between experts;

(2) inheritance networks do not allow us to reason with conjunctive/disjunctive knowledge: for example, the statement "If Clyde is an elephant and Clyde has big ears, then Clyde is an African elephant" is not expressed naturally in terms of inheritance nets. Until recently, inheritance nets have not been characterised with convenient declarative semantics. Authors in [12] rectified this situation. They also developed a theory of reasoning by inheritance based on Subrahmanian annotated logic.

12.1.4 Object Oriented Database

An essential use of database data is to combine datasets. For example, we might want to know the names of everyone in a company which earns a salary above \$1,000. In this case, the manipulated object is a set. More complex operations can be performed based on our knowledge of this set. For every individual earning more than US\$1,000, we may want to obtain additional data, such as the college he graduated from, his age, etc. We then have a set in which each element is a structured object containing information about faculties, ages, etc. The conservation of such base data has been highly complex, as inconsistencies about these sets can arise in the strangest ways. Many thought no object-oriented database could interact or deductive databases [14]. One reason for this view was these strange inconsistencies. Authors in [11] showed that paraconsistent logic elegantly provides a rationale for object-oriented databases via annotated logic.

12.2 Some Subsequent Applications

12.2.1 Logic Programming

The Annotated logic has a remarkable appearance. Its composition as two-sorted logic in which one of the variables has an ordered structure has found fertile ground in applications. Let us comment on some aspects of these applications.

In [2, 6] it was developed a Paraconsistent logic programming language—Paralog. As it is well known, the development of computationally efficient programs in it should exploit the following two aspects in its language:

1. The declarative aspect that describes the logical structure of the problem, and
2. The procedural aspect that describes how the computer solves the problem. However, it is not always an easy task to conciliate both aspects. Therefore, programs to be implemented in Paralog should be well defined to evidence both the declarative and procedural aspects of the language.

It must be pointed out that programs in Paralog, like programs in standard Prolog, may be easily understood or reduced—when well defined—by adding or eliminating clauses, respectively.

It is worth noting that the implementation was based on the references mentioned above but independently.

12.2.2 Paraconsistent Annotated Evidential Logic Eτ

We can observe that the lattice associated with the Annotated logic is arbitrary but fixed. Depending on each application, the lattice plays a crucial role in reflecting the desired properties.

One of the most interesting lattices is the following: we consider the ordered system $\tau = <[0, 1]\times[0, 1], \leq>$ where $\leq$ is the order relation defined by $(\mu_1, \lambda_1) \leq (\mu_2, \lambda_2) \leftrightarrow \mu_1 \leq \mu_2$ and $\lambda_2 \leq \lambda_1$, where [0, 1] is the closed unitary real number with the usual order.

When considering the lattice τ above, the correspondent logic is called Paraconsistent annotated evidential logic Eτ—Logic Eτ. The basic formulas of the logic Eτ are $p(\mu, \lambda)$, where $(\mu, \lambda) \in [0, 1]^2$ and [0, 1] is the real unitary interval (p denotes a propositional variable). $p(\mu, \lambda)$ can be read (among other readings): "It is assumed that p's favourable evidence is μ and contrary evidence is λ."

Thus,

- p(1.0, 0.0) can be interpreted as a true proposition,
- p(0.0, 1.0) as false proposition,
- p(1.0, 1.0) as inconsistent proposition,
- p(0.0, 0.0) as paracomplete proposition, and
- p(0.5, 0.5) as an indefinite proposition. Also, we introduce the following concepts:
- Uncertainty degree: $Gun(\mu, \lambda) = \mu + \lambda - 1$ $(0 \leq \mu, \lambda \leq 1)$;
- Certainty degree: $Gce(\mu, \lambda) = \mu - \lambda$ $(0 \leq \mu, \lambda \leq 1)$;

With the uncertainty and certainty degrees, we can get the following 12 output states: extreme and non-extreme, as shown in Table 12.1.

All states are represented in the following Fig. 12.1.

Some additional control values are:

V_{cic} = maximum value of uncertainty control = C3.

V_{cve} = maximum value of certainty control = C1.

V_{cpa} = minimum value of uncertainty control = C4.

V_{cfa} = minimum value of certainty control = C2.

We have an output in each input's decision lattice of favourable and contrary evidence. These steps can be implemented in a simple algorithm (Para-analyser) and hardware (Para-control) which we will see later [7]. One of the applications of the algorithm is in expert systems, which we will show in some topics (Fig. 12.2).

Table 12.1 Extreme and Non-extreme states

Extreme States	**Symbol**
True	V
False	F
Inconsistent	T
Paracomplete	⊥
Non-extreme states	**Symbol**
Quasi-true tending to Inconsistent	QV→T
Quasi-true tending to Paracomplete	QV→⊥
Quasi-false tending to Inconsistent	QF→T
Quasi-false tending to Paracomplete	QF→⊥
Quasi-inconsistent tending to True	QT→V
Quasi-inconsistent tending to False	QT→F
Quasi-paracomplete tending to True	Q⊥→V
Quasi-paracomplete tending to False	Q⊥→F

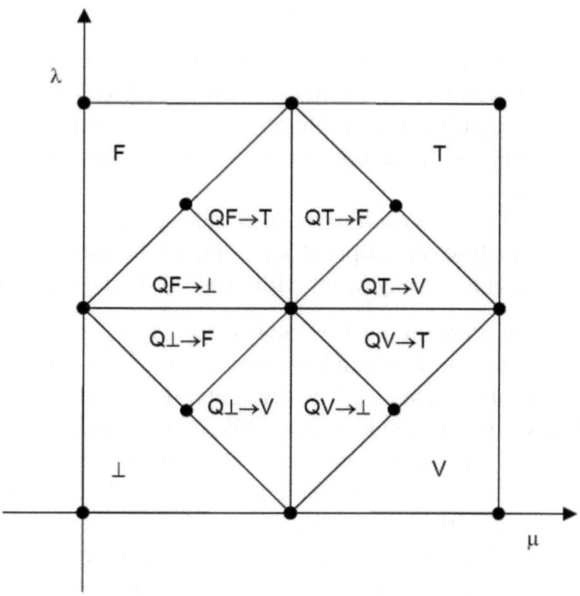

Fig. 12.1 Lattice τ

12.2.3 Expert Systems

Next, we discuss software based on logic Eτ (that plays the role of an expert system) using expert knowledge databases to offer advice or make decisions. Such a system is based on the language of logic Eτ, presenting certain peculiarities that we have already commented on. The expert system is as follows:

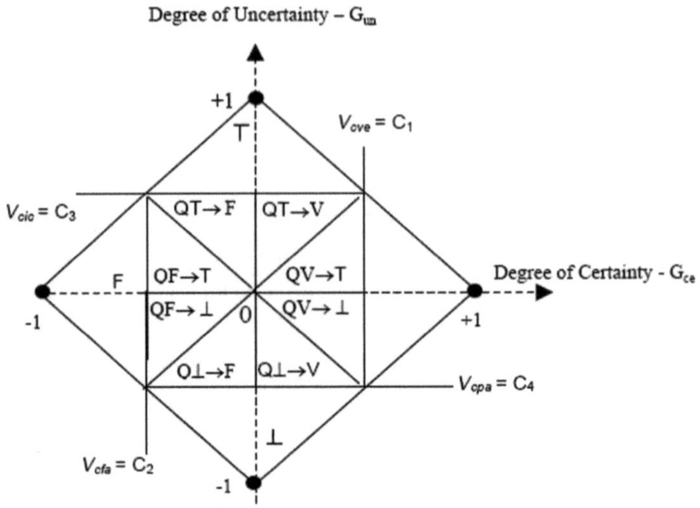

Fig. 12.2 Certainty and uncertainty degrees

(1) Fix a statement (proposition) about the problem under consideration—for example, 'On Avenida 5, corner of Avenida 6 is a good place to open a cafe'.
(2) Regarding the problem, we started considering all the parameters necessary to analyse it, such as location, possible customers, if it is close to a large office, etc.
(3) Once the data have been obtained, it is necessary to list the specialists who will analyse the proposal according to the chosen parameters. In this case, 'expert' can mean any agent with expertise in the topic or objective quantitative data.

The problem must be analysed in all aspects necessary for a complete analysis. Thus, it is desirable to have several groups of experts who will analyse a parameter from different angles. The parameters can have different importance. Each parameter is assigned a weight to map this feature. Finally, the idea is that several groups of experts analyse the parameter globally. Another important observation is that each specialist is different from the others; i.e., they may know a greater or lower degree. Thus, experts can also be given different weights. Note that, unlike the data from a joint knowledge base, the knowledge bases using logic Eτ, each expert issues favourable and contrary evidence for each assertion he analyses. In this way, the language of the logic Eτ is more faithful in describing the portion of reality under consideration and can lead to a more realistic analysis.

Next, we illustrate how the evidence (favourable and unfavorable) presented by experts can be combined. Details of each edition the reader can find in reference [1] (Fig. 12.3).

Next, we present some applications of the expert system.

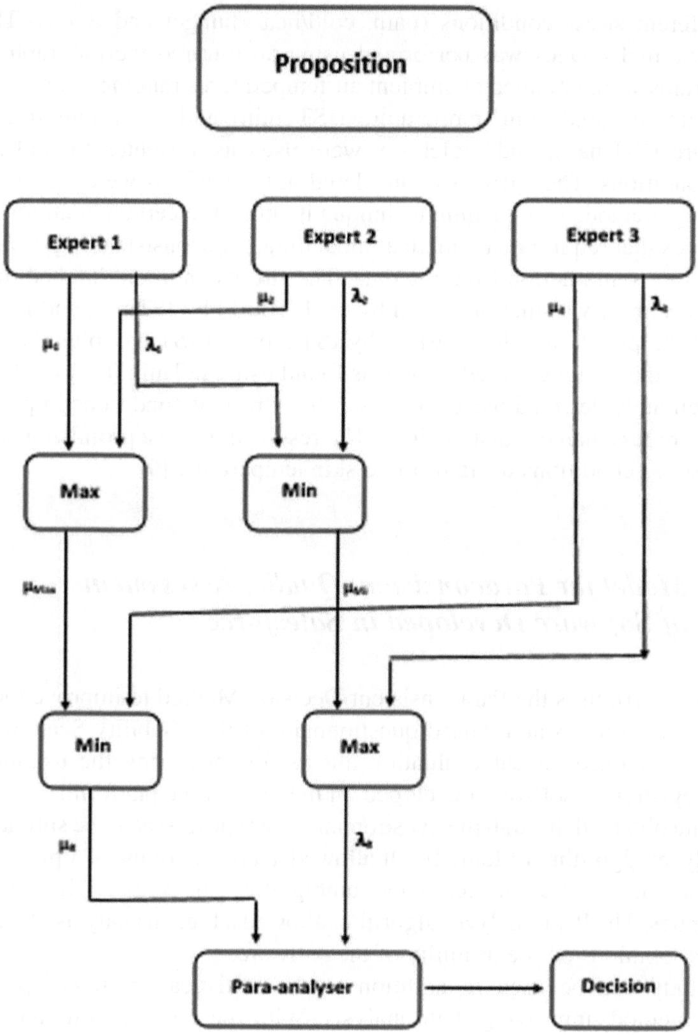

Fig. 12.3 Scheme of an expert system based on logic Eτ and Para-analyzer algorithm

12.2.4 Automatic Prediction of Stress in Piglets (Sus Scrofa)

Consumption of pork grows at around 5% per year in developing countries. Ensuring food safety within the ethical standards of meat production is a growing consumer demand. The study in [9] aimed to develop a model to predict piglet stress based on infrared skin temperature (IST) using machine learning and paraconsistent logic. Seventy-two piglets (32 males and 40 females) from 1 to 52 days of age had their infrared skin temperature recorded during farrowing and nursery phases under different stress conditions (pain, cold/heat, hunger and thirst). The evaluation of thermal images was performed using an infrared thermography camera. Thermograms were obtained at ambient air temperatures ranging from 24 to 30 °C.

Minimum infrared skin temperature (IST min) and maximum infrared skin temperature (ISTmax), and piglet sex were used as attributes to find the stress (target) conditions. The attributes considered in the analysis were classified by the data mining method. The imaging technique is subject to certain contradictions and uncertainties that require mathematical modelling. Paraconsistent logic was applied to extract the contradiction from the data. The stress condition that had the highest detection accuracy was that predicted by cold (100%) by ISTmin, and ISTmin plus the sex of the piglet, and thirst (91%) by ISTmax and ISTmax plus the sex of the piglet. The highest hunger prediction was found using ISTmin (86%). Although the model accurately detected these stresses, other stressful conditions in piglets, such as pain, were less than or equal to 50%. The results indicate a promising assessment of piglet stress conditions using infrared skin temperature [9].

12.2.5 Model for Paraconsistent Quality Assessment of Software Developed in Salesforce

The study in [10] uses the Paraconsistent Decision Method to improve the analysis of data captured in a standardised questionnaire of the Usability Scale of the SUS System. The paraconsistent evaluation allows for measuring the usability of the supplier registration software developed on the SalesForce platform.

The data obtained through the questionnaire were processed to be submitted to the Para-analyzer algorithm of logic Eτ. It allowed analysing of users' opinions considering their uncertainties, inaccuracies, ambiguities and subjectivities inherent to human values. The Para-analyzer algorithm allowed a logical analysis of a consensus of experts' opinions on the usability of the software.

The study can be used in addition to the statistical treatment provided by the SUS method, improving data analysis. With the result of this analysis, it is possible to diagnose usability problems, contributing to the improvement of software development.

12.2.6 About the Turning Point of Cache Efficiency in Computer Networks with Logic Eτ

Object caches minimise data traffic in many areas of Information Technology, including computer networks. In this scenario, they are usually hosted in proxies, storing page objects (texts, pictures, among others) and implementing access control policies. Its correct operation can provide a significant performance gain in data exchange, allowing immediate response to requested resources. In [4], we investigated the different efficiency states of two distinct types of computer network caches and determined the dynamics of change of these states with logic Eτ.

12.2.7 Robotics

The Paracontrol is the electric-electronic materialisation of the Para-analyzer algorithm [7], which is an electronic circuit (logical controller) which treats logical signals in the context of logic Eτ. Such a circuit compares logical values and determines domains of a state lattice corresponding to the output value. A voltage represents favourable evidence and contrary evidence degrees. Analog of operational amplifiers determine Certainty and Uncertainty degrees.

The Paracontrol comprises analogue and digital systems and can be externally adjusted by applying positive and negative voltages. The Paracontrol was tested in real-life experiments with an autonomous mobile robot Emmy, whose favourable/contrary evidences coincide with the values of ultrasonic sensors and distances are represented by continuous values of voltage (Fig. 12.4).

The Paracontrol controller has been applied to a series of autonomous mobile robots. In some previous works [7], the Emmy autonomous mobile robot is presented.

When moving in an unstructured environment, the robot Emmy obtains information about the presence/absence of obstacles through the sonar system called

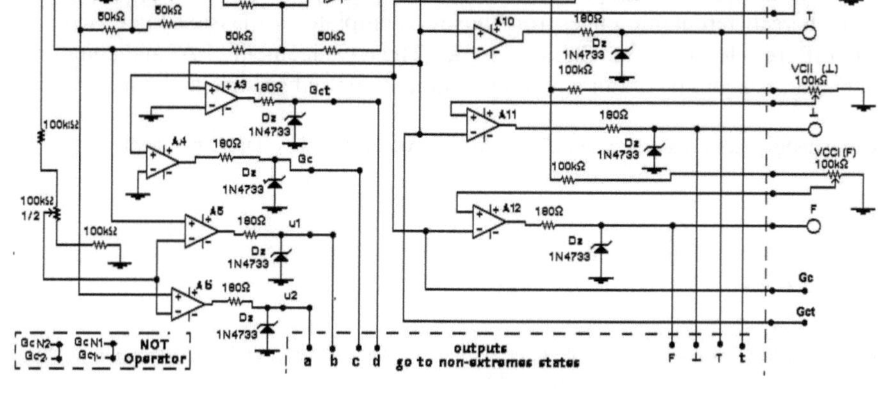

Fig. 12.4 Paracontrol circuit

Parasonic [1, 7]. The Parasonic is made up of two POLAROID 6500 ultrasonic sensors controlled by an 8051 microcontroller. The sensors detect obstacles on the way, transforming the distances to the obstacle into electrical signals of continuous voltage ranging from 0 to 5 volts.

The Emmy robot uses the paracontrol system to travel in unstructured environments, avoiding collisions with human beings, objects, walls, tables, etc. The reception of information about obstacles is called non-contact, which is the method to obtain and treat signals from ultrasonic or optical sensors to avoid collisions.

The work showed that applying Paraconsistent Annotated Evidential Logic Eτ in algorithms was practical and could be done directly without extralogical devices, even in conflicts or lack of information. In this way, it expands the contribution of several robotic navigation systems for decision-making. The paraconsistent algorithm developed proved efficient and easy to integrate with other systems. In addition, using a servo motor to control the robot's direction ensures agility, speed and precision in movements. As future work, the Paraconsistent logic can be extended to other uses in navigation systems, such as positioning on the map, among other topics.

12.3 Conclusion

For nearly two thousand years, classical logic dominated human thought. However, with the advent mainly of computational applications in AI, Automation, Robotics, and other branches of knowledge, new inference automation was necessary, and consequently, new logical systems were being created. Thus emerged, to name a few, the theory of Fuzzy sets, the theory of Rough sets, non-monotonic logic, defeasible logic, linear logic, quantum logic, etc. Paraconsistent logic has found, in recent decades, critical applications in applied science. It is somewhat natural, as inconsistencies naturally appear in applications. Some examples include big data-bases, diagnostics, and pattern detection where there is inherent ambiguity, conflicting information, etc. The advent of systems different from classical logic raises questions such as: are there different inferences from those contemplated in classical logic, that is, are there different logics from Classical logic? The advancement of AI and computing, in general, is bringing about a real revolution in applied science and pure science.

Acknowledgements I am very grateful to Prof. S. Akama for his valuable comments.

References

1. Abe, J.M.: Paraconsistent intelligent-based systems: New trends in the applications of paraconsistency, vol. 94, Springer, 2015
2. Abe, J.M., Akama, S., Nakamatsu, K.: Introduction to annotated logics—foundations for paracomplete and paraconsistent reasoning, series title intelligent systems reference library, Volume

88. Springer International Publishing, Copyright Holder Springer International Publishing Switzerland, ISBN 978-3-319-17911-7, Edition Number 1, 190 pages, 2015
3. Akama, S.: Towards paraconsistent engineering, intelligent systems reference library, Volume 110, 234 pages, ISBN: 978–3–319–40417–2 (Print) 978–3–319–40418–9 (Online), Series ISSN 1868–4394, Publisher Springer International Publishing, DOI: https://doi.org/10.1007/978-3-319-40418-9 (2016)
4. Blair, H. A., Subrahmanian, V. S.: Paraconsistent logic programming. In: Proc. 7th conference on foundations of software technology and theoretical computer science, lecture notes in computer science, vol. 287, 340–360, Springer-Verlag (1987)
5. Pimenta Jr, A., Minoro Abe, J.:Determination of the turning point of cache efficiency in computer networks with logic Eτ. In: Procedia Computer Science, Elsevier, ISSN: 1877-0509, Procedia Computer Science, Vol. 159, pp. 1182–1189 (2019). https://doi.org/10.1016/j.procs.2019.09.287
6. da Costa, N.C.A., Prado, J.P.A., Abe, J.M., Ávila, B.C, Rillo, M.: Paralog: Um Prolog Paraconsistente baseado em Lógica Anotada, Coleção Documentos, Série Lógica e Teoria da Ciência, IEA-USP, no 18, ISSN 16799429, p. 21 (1995)
7. da Silva Filho, J.I.: Métodos de interpretação da Lógica Paraconsistente Anotada com anotação com dois valores LPA2v com construção de Algoritmo e implementação de Circuitos Eletrônicos, EPUSP, PhD Thesis (in Portuguese), São Paulo (1999)
8. de Carvalho, Fábio R.; Abe, Jair M. A Paraconsistent Decision-Making Method. Smart Innovation, Systems and Technologies 87. Springer International Publishing; ISBN 978-3-319-74109-3; ISBN eBook 978-3-319-74110-9; https://doi.org/10.1007/978-3-319-74110-9. Switzerland, p. 212 (2018)
9. Fonseca, F., Abe, J.M., Nääs, I.A., Cordeiro, A.F., Amaral, F., Ungaro, H.: Automatic prediction of stress in piglets (Sus Scrofa) using infrared skin temperature. Comput. Electron. Agric. 168, 105148, 1–11 (2020). https://doi.org/10.1016/j.compag.2019.105148
10. Forçan L.R., Abe J.M., de Lima L.A., Nascimento S.S.: Questionnaire model for paraconsistent quality assessment of software developed in SalesForce. In: Lalic B., Majstorovic V., Marjanovic U., von Cieminski G., Romero D. (eds) Advances in production management systems. The path to digital transformation and innovation of production management systems. APMS 2020. IFIP Advances in Information and Communication Technology, vol 591. Springer, Cham (2020). https://doi.org/10.1007/978-3030-57993-7_38
11. Kifer, M., Wu, J.: A logic for Object-Oriented Logic Programming. In: Proc. 8th ACM Symp. on Principles of Database Systems, pp. 379–393 (1989)
12. Kifer, M., Krishnaprasad, T.: An evidence-based for a theory of inheritance. In: Proc. 11th International Joint Conf. on Artificial Intelligence, 1093–1098, Morgan-Kaufmann (1989)
13. Sylvan, R., Abe, J.M.: On general annotated logics, with an introduction to full accounting logics. Bull. Symb. Logic **2**, 118–119 (1996)
14. Ullman,L.A.: Database Theory: Past and Future, Proc. of the ACM SIGIGACT/SI/SIGMOD Symp. on Principles of Database Systems, pp. 1–10 (1987)

MIX
Papier aus verantwortungsvollen Quellen
Paper from responsible sources
FSC® C105338

If you have any concerns about our products,
you can contact us on
ProductSafety@springernature.com

In case Publisher is established outside the EU,
the EU authorized representative is:
Springer Nature Customer Service Center GmbH
Europaplatz 3, 69115 Heidelberg, Germany

Printed by Libri Plureos GmbH
in Hamburg, Germany